ORCHIDS OF INDIA

COMMERCIALIZATION AND CONSERVATION

ORCHIDS OF INDIA

COMMERCIALIZATION AND CONSERVATION

Jauti Sharma

2021

Daya Publishing House®

A Division of

Astral International Pvt. Ltd.

New Delhi - 110 002

First Published, 1996
Reprinted, 2021

ISBN: 978-81-7035-153-5 (Hardbound)

Published by : **Daya Publishing House®**
A Division of
Astral International Pvt. Ltd.
– ISO 9001:2015 Certified Company –
4736/23, Ansari Road, Darya Ganj
New Delhi-110 002
Ph. 011-43549197, 23278134
E-mail: info@astralint.com
Website: www.astralint.com

Digitally Printed at : **Replika Press Pvt. Ltd.**

ACKNOWLEDGEMENTS

This is to acknowledge my deep sense of gratitude to my supervisor, Professor (Dr.) Y.S. Chauhan, Department of Botany, School of Life Sciences, North Eastern Hill University, Shillong for his encouragement, keen interest, valuable suggestions and constructive criticisms for preparing and editing my research work.

My grateful thanks to the Heads of Botany Department N.E.H.U., Shillong for facilities provided.

I am grateful to Professor (Dr.) R.S. Tripathi, Dr. R.P. Singh for their help and suggestions. My grateful thanks are also due to the former Head of the Botany Department, Professor (Dr.) P.S. Ramakrishnan, FNA, for suggesting the right research problem.

Professor (Dr.) R. George Michael, Professor (Late) Dr. M.K. Khare, of Zoology Deptt., and Professor (Dr.) D.T. Khathing, Head, R.S.I.C., from N.E.H.U., Shillong, were a constant source in providing mental strength in the completion of this work, my sincere thanks to them also.

I am grateful to Fr. S. Mavely, Principal, St. Anthony's College, Shillong for his words of encouragements to continue with my research work.

I am grateful to Mr. George Middle Cot, A real Orchidologist for his constant encouragements and fruitful discussions. Mr. Hariprasad of Light Green Nursery, Shillong was always very kind and helpful and allowed me to conduct some of the field experiments. To Dr. (Mrs.) A.R. Tripathi, I owe special words of thanks.

My warmest and sincere thanks are also due to R.S.W.A., Dr. (Miss) M. Lyngdoh, Dr. Vituo Belho, Dr. T.S. Rathore, Dr. S.C. Joshi, Dr. A.N. Verma, Dr. A.L.S. Rajee, Dr. S.R. Katiyar, Dr. S.K. Sharma, Dr. (Ms.) Mamtaz Dkhar, Dr. (Mrs.) Nirja Mishra, Dr. (Mrs.) A. Khan, Mr. S.K. Roy, Dr. P.B. Gurung, Dr. D. Dutta, Mr. V.R. Rengsi, Mr. Amit Biswas for their help and co-operation. Dr. (Mrs.) Tejimala Gurung Nag proved a real friend in need. I mut acknowledge the immense help rendered by Dr. S. Changkija and Mr. K.A. Joy. My special thanks are also to Mr. K.A. Joy for his relentless and constant help for the draft as well as final typing.

I must mention a special word of thanks to Director, I.C.A.R. (NE region) and Deputy Director, B.S.I. (EC), Shillong for Library facilities. My thanks are also due to Sri B.K. Das for photography and to all Office, Store and Technical Staff of Botany Deptt., N.E.H.U., Shillong for their co-operation during the present investigation.

Lastly, I will always remain indebted to my parents and brothers who are truly the source of inspiration to complete this research work and I dedicate this work to them.

JAUTI SHARMA

CONTENTS

1
Introduction

Out of 20,000 orchid species world over, India accounts for over a thousand species of epiphytic, terrestrial and saprophytic orchids (Jain and Mehrotra, 1984; Karthikeyan *et al.*, 1989; Chadha, 1992). Majority of Indian orchids (about 700 species) are found in phytogeographically interesting northeastern region. Based on phytogeographical evidences, India is considered primary/secondary centre of origin for orchids (Jain, 1986). This region is not only rich in number of orchid taxa which grow profusely in nature, but most species are at top of the list of ornamental orchids.

However, this natural orchid wealth is yet to be judiciously utilized and managed. There is always the danger that unless scientific conservation measures are taken along with proper utilization, India in general and northeast in particular is sure to loose most of the valuable orchid wealth that are of medicinal, horticultural and aesthetic values (Rao, 1979; Arora and Mukherjee, 1983; Handa, 1986). Several orchid species have become extinct, many are endangered and several are at the verge of extinction due to various activities e. g., indiscriminate cutting of forest, shifting agriculture and ruthless exploitation of orchids for trade. Many of the Indian orchids particularly the species of *Cymbididium, Dendrobium, Paphiopedilum, Renanthera, Vanda* are in International horticultural trade. The orchids are habitat specific and their trade from naturally grown population has been banned by Government of India. The clandestine trade, however, is continuing without any check. Annually, the U.S.A. alone imports 50,000 plants of Lady's Slipper orchids and the number is increasing every year and nearly 70 per cent of these plants are illegally exported from tropical countries like India, Thailand etc., (Chadha, 1992). *The desire for new and exotic colours and the demand* in the Japanese market for orchids with yellow and golden flowers bode well for the Indian orchids. According to Alvensleben (1980), *future sales of tropical orchids are likely to increase.* Countries like Thailand, Singapore, Malaysia, Japan, Korea, Taiwan are now running multimillion dollar business in orchids using tissue culture techniques (Gavinlertvatana and Prutpongse, 1991). However, commercial orchid growing in India, is yet to be properly organized, and the trade is virtually in the hands of some dealers who collect orchids from wild populations to meet their foreign and local commitments.

The orchid seeds are minute and produced in million per capsule. The seeds however, lack metabolic machinery and functional endosperm with the result that only 0.2 – 0.3 per cent of the seeds germinate in nature (Singh, 1992). Bernard (1899) discovered that in nature orchid seeds will germinate only if infected by a fungus. The seed germination of orchids on mycorrhizal association was first described by Bernard in 1909. However, the seed take long period for their germination and any disturbance to the habitat or physical environment destroys the whole population. Knudson (1922) achieved success in asymbiotic germination of orchid seeds for the first time on sugar enriched mineral medium using tissue culture technique. Later on success in the germination and seedling formation on Knudson's asymbiotic medium was reported by Anon. (1924, 1925 a, b, c); Ballion and Ballion (1924, 1928), Clement (1925) and Smith (1932). Knudson (1946, 1950, 1951) following his initial work devoted attention into the development of better culture medium. Many investigators attempted to improve Knudson's culture medium according to requirements of various species (Robinson, 1941; Sampolinski, 1965). Attempts to germinate seeds of terrestrial orchids have only been partially successful (Van Waes and Debergh, 1986). In some species the seed coat may be impregnated with inhibiting substances, capable of slowing down germination (Withner, 1955).

Arditti (1961) reported that orchid seed germination resembles the development of a dormant bud more than it resembles a germinating seed. Chlorophyll may or may not be formed following the initial swelling of the embryo. The germinating embryo swells further and bursts out of the seed coat, and a cone-shaped or spherical seedling is soon formed. This is called protocorm stage (Bernard, 1909). Soon the first miniature leaves appear followed by the appearance of first root, resulting in a slowly developing miniature plant (Arditti, 1961). With minor modifications germination in most orchids proceeds as described above (Knudson, 1922; Carlson, 1935; Burgeff, 1936, 1959; Curtis, 1943; Poddubnaya-Arnoldi and Selezneva, 1957; Arditti and Ernst, 1984). The first germinating orchid seeds to be described were apparently those of *Bletilla* Rchb. f., *Orchis morio* L., and *Limodorum verecundum* Prodr. (Salisbury, 1804; Burgeff, 1909).

The *in vitro* techniques provide better understanding of different physico-chemical requirement that affect growth and may even respond differently at various stages of growth (Arditti *et al.*, 1981). Selection of right age of the capsule has an important role for early and increased seed germination besides the other physico-chemical factors.

Although most orchid seeds can germinate and grow readily on a number of media (Arditti, 1967; Withner, 1959), some genera are more recalcitrant. Diandrous orchids like *Paphiopedilum*, *Cypripedium*, *Phragmipedium* and *Selenipedium* have proved difficult to propagate from seed (Ernst, 1974). Epiphytic and terrestrial species differ greatly in their requirements for mineral nutrients, nitrogenous sources, light and pH for their germination and growth (Arditti and Ernst, 1984). The essentiality of exogenous carbohydrates in germination of orchid seed is greatly emphasized. The most commonly used sugar is sucrose (Arditti and Ernst, 1984). Addition of growth hormones and organic supplements are also the need in some species for the enhancement of seed germination and seedling growth. Thus, study of, selection of right age of capsule and physico-chemical requirement becomes important for rapid mass multiplication of horticulturally important orchid species. Arditti *et al.*, (1982) emphasized that in orchids the above requirements vary with the genus, species and locality. Therefore, in the present study an attempt is made to investigate these aspects in the lesser worked out *Dendrobium* (epiphytic) and *Paphiopedilum* (terrestrial) species growing in the N.E. region.

Orchids being highly heterozygous it is difficult to multiply the required mother plant through seed culture. Asexual propagation is the only means to obtain a large number of selected recombinant types in the out breeding taxa of Orchidaceae (Wilfret, 1966). Vegetative propagation is very slow (Singh, 1976). The technique of meristem tissue culture makes it possible to multiply unlimited numbers of any desirable clone. Plant tissue culture or the aseptic culture of cells, tissues and organs, is an important tool in basic and applied studies having commercial application (Thorpe, 1990). The technology owes its origin to the ideas of the German plant physiologist, Haberlandt, who in his famous address to the German Academy in 1902 suggested it should be possible to cultivate artificial embryos from vegetative cells. He introduced the concept of totipotency: that all living cells containing a normal complement of chromosomes should be capable of regenerating entire plant (Krikorian and Berquam, 1969).

The earliest report of using tissue culture technique in orchids goes back to 1949 when it was demonstrated that *Phalaenopsis* plantlets could develop from the buds of inflorescence stalks (Rotor, 1949). However, the credit of achieving mass clonal propagation of orchids goes to Morel (1960) who was successful in culturing *Cymbidium* shoot apices on Knudson C agar medium. The explant first became green and then enlarged slowly as a small bulblet similar to a protocorm-like body (Morel, 1960). Morel (1964) reported that in a year more than 4 million plant from a single explant can be obtained. Following the technique of Morel (1964) further success in other species with shoot tips are reported in *Aranda* (Goh, 1973; Goh *et al.*, 1975); *Aranthera* (Irawati *et al.*, 1977); *Rhynchostylis* (Vajrabhaya and Vajrabhaya, 1970); *Dendrobium* (Kim *et al.*, 1970; Morel, 1974; Singh, 1976; Sagawa and Kunisaki, 1984); *Vanda* (Goh, 1970). Since Morel's work in 1960, shoot tip remained most commonly used explants (Wimber, 1963; Wilfret, 1966; Steward and Mapes, 1971; Werckmeister, 1971; Fonnesbech, 1972; Kim *et al.*, 1970; Morel, 1974; Sagawa and Kunisaki, 1984; Sharon and Vasundhara, 1990). Other plant parts like leaves (Champagnat *et al.*, 1970; Loh *et al.*, 1975; Zimmer and Peiper, 1975; Vij *et al.*, 1984; Chaturvedi and Sharma, 1986; Seeni, 1988; Vij and Pathak, 1990), axillary buds (Kim *et al.*, 1970; Goh, 1973; Goh and Loh, 1975; Singh, 1976; Goh, 1981) inflorescences (Tse *et al.*, 1971; Intuwong *et al.*, 1971; Intuwong and Sagawa, 1973) and roots (Sood and Vij, 1986) for clonal propagation has also been used.

In a developing system hormones remain in a very intricate pattern with the fluctuation of their compounds growth regulation takes place. Most of the media used for orchid seed germination or tissue culture are same as those devised for other plant tissues (Withner, 1959; Arditti, 1967). Source of the explants, size of the explants, pH and other environmental factors have also been reported to be very important in mass clonal propagation (Bertsch, 1966; Marston, 1966; Marston and Voraurai, 1967, 1969; Goh, 1971).

Haber (1968) emphasized the importance of ionizing radiations as a research tool. Modifications of various growth and development of various plant parts, both vegetative and reproductive have been induced by radiation (Gunckel and Sparrow, 1961).

Conservation of biodiversity is a fundamental requirement of plant breeding, horticulture and biotechnology (Holden and Williams, 1984; Wilkes, 1983). Maintaining a genotype in pots is not only risky but also expensive and time consuming (Bhojwani, 1981). *In vitro* storage is one component of an alternative conservation strategy which embraces propagation and movement of germplasm (Thorpe, 1990).

The last and the most important step is the establishment of the *in vitro* raised seedlings in field conditions, so that they can be grown in nature and for meeting the market demand. This would save orchids from getting exploited from nature.

In view of the above studies effect of capsule age was investigated in ornamentally important Dendrobiums (*D. aphyllum* (Roxb.) Fischer, *D. chrysanthum* Lindl., *D. chrysotoxum* Lindl., *D. falconeri* Hook., *D. heterocarpum* Wall. ex Lindl., *D. ochreatum* Lindl., *D. primulianum* Lindl., *D. wardianum* Warner) (Plate 1a. 1-8) and Paphiopedilums (*P. fairrieanum* (Lindl.) Stein, *P. hirsutissimum* (Lindl. ex hook.) Stein, *P. insigne* (Wall. ex Lindl.) Pfitz., *P. spicerianum* (Reichb. f.) Pfitz., *P. venustum* (Sims) Pfitz. and *P. villosum* (Lindl.) Stein) (Plate 1b. 1-6) growing in the northeastern region. Considering the importance of *D. chrysanthum* and *P. spicerianum* in horticulture more emphasis was given to these species by investigating the following additional aspects so that a package of practice could be devised for optimum seed germination, rapid multiplication and conservations of these species:

1. Effect of different nutrient media on the seed germination and seedling growth.
2. Effect of light and pH on seed germination and seedling growth.
3. Effect of carbohydrates, nitrogen sources, amino acids, growth hormones and organic supplements on seed germination and seedling growth.
4. Effect of different media and growth hormones on meristem culture.
5. Cold preservation: Low temperature requirement for storage of protocorms.
6. Establishment of *in vitro* raised seedling in field conditions.

Plate 6a : Shoot formation in shoot tip and root tip explants of *D. chrysanthum*

1. formation of protocorm like bodies at the base of shoot tips on Kn C medium containing 15% coconut milk, 60 days after culture
2. differentiation of shoots from the protocorm-like bodies on Kn C medium containing 15% coconut milk, 90 days after culture
3. formation of shoots (4-5 mm) from root tip culture on Kn C medium containing 15% coconut milk and 0.1% activated charcoal, 180 days after culture

1
2
3

Plate 1 b : Flowers of *Paphiopedilum* species

1. *P. fairrieanum*
2. *P. hirsutissimum*
3. *P. insigne*
4. *P. spicerianum*
5. *P. venustum*
6. *P. villosum*

1

2

3

4

5

6

2
Effect of Capsule Age on Seed Germination

Introduction

In the natural conditions majority of orchid flowers are not pollinated and capsules are rarely formed (Rao, 1977). The number of seeds produced per capsule varies from 5,000- 6,000 in *Cephalanthera* to 2 to 3 million in *Cynoches* and *Cattleya* (Salisbury, 1942; Shushan, 1959; Withner, 1959; Jeyanayaghy and Rao, 1966; Arditti, 1967). The seeds are non-endospermic. In the natural environment, less than 5% seeds germinate (Rao, 1977). Prior to 1922, it was considered that an orchid seed would germinate and continue growing only if it had been invaded by the proper endotrophic mycorrhiza (Bernard, 1904). In 1922, Knudson demonstrated that it is possible to germinate orchid seeds under aseptic conditions using sugar enriched mineral media. Borris (1969), Vermeulen (1947) and Fast (1974) observed that seed from the immature pods can also be germinated. Oliva and Arditti (1984) compared seed germination in seeds of immature/undehisced capsule and seeds from dehisced capsule using *in vitro* seed germination method. Seeds of dehisced capsules reveal poor germination compared to seeds from undehisced capsule (Arditti *et al.*, 1982). The stage of maturation of seed also influenced the rate of *in vitro* germination (Linden, 1980).

In most of the species belonging to Monandrae group, initiation of division of cells to form the archesporial tissue depends upon the stimulatory effect of auxin from pollen and does not usually begin until pollination takes place (Wirth and Withner, 1959). In the diandrous species, however, development of embryo sac reaches an advanced stage before pollination (Duncan and Curtis, 1942). Ovule formation was reported to occur 43 to 45 days after pollination in *Cymbidium* and 58 to 60 days after pollination in *Paphiopedilum*. For maturation of embryo, 115 to 120 days and 195 to 200 days after pollination were required, respectively in *Cymbidium* and *Paphiopedilum* (Nagashima, 1982). The time gap between pollination and fertilisation is 60-75 days in *Dendrobium* spp. (Swamy, 1949), 150 days in *Paphiopedilum venustum* (Treub, 1879) and 105 days in *Paphiopedilum villosum* (Duncan and Curtis, 1942).

The earliest stage at which embryo can be cultured successfully varies with the genus, species, hybrid and local conditions (Arditti *et al.*, 1982). Since these informations are not available for some of the horticulturally important *Dendrobium* species (*D. aphyllum, D. chrysanthum, D. chrysotoxum, D. falconeri, D. heterocarpum, D. ochreatum, D. primulianum, D. wardianum*) and *Paphiopedilum* species (*P. fairrieanum, P. hirsutissimum, P. insigne, P. spicerianum, P. venustum, P. villosum*) in the present study, an attempt was made to investigate these aspects. The characters considered were capsule age and its bearing on seed germinability, time taken to commence germination and percentage of germination.

MATERIALS AND METHODS

The most appropriate stage of capsule for optimal seed germination was determined in eight species of *Dendrobium* and six species of *Paphiopedilum* mentioned above. For the purpose experimental plants were identified in a local nursery (Light Green Nursery, Upper Shillong), and 15 plants of each species growing almost under natural conditions were selected. For each species, atleast 15 flowers were sib-mated and capsules thus formed were harvested at different ages (measured in days after pollination) viz., 90 days, 150 days, 210 days, 270 days, 330 days and 360 days. For every collection two capsules were selected. During this period the experimental plants were kept under constant observations. Kn C (Knudson C medium, 1946) and Th GD (Thomale GD medium, 1954) medium were used for *Dendrobium* and *Paphiopedilum* respectively. The pH of the media was always adjusted to 5.0. 15.0 ml of medium was taken in 25 × 125 mm borosil test tubes. The tube mouth was closed with cotton plugs. The tubes were autoclaved at 121°C and 15 lb/in^2 pressure for 20 minutes.

Seeds were cultured on the respective media under aseptic conditions. Cultures of *Dendrobium* were kept under continuous light (2,500-3,000 lux) at 25 ± 1°C and *Paphiopedilum* were kept under dark at 25 ± 1°C because seeds of *Paphiopedilum* do not germinate in light.

Bursting of seed coat and emergence of the enlarged embryos, i.e., the protocorms was considered germination. The germination of seeds was recorded and percentage seed germination calculated. Time taken for the commencement of germination was observed at 5 days interval.

Detailed Method of Sampling and Counting

Five samples of the seeds from two tubes each were taken out after 30 days in case of *Dendrobium* and 45 days in case of *Paphiopedilum*. Each sample was considered a replicate and 10 slides were prepared for observations by placing one sample in a drop of glycerine and covering it with a cover slip. Per slide 20 seeds were scrutinized randomly under a compound microscope. The seeds were classified germinated/ungerminated for calculating germination percentage. Per species a total of 200 seeds were scrutinised.

The mean and standard errors (S.E.) of 10 replicates were determined and students 't' test was performed for determining the significance.

RESULTS

Dendrobium: Capsules of Dendrobiums remain green and dehisce to disperse the seeds any time beyond 270 days in case of *D. aphyllum, D. falconeri;* 330 days in case of *D. primulianum, D. wardianum;* and 360 days in case of *D. chrysanthum, D. chrysotoxum, D. heterocarpum* and *D. ochreatum* (Table 2.1).

Table 2.1: Effect of capsule age on seed germination in *Dendrobium* spp. after 30 days

Species	*Capsule Age (Days)*	*Time to Commence Germination (Days)*	*Percentage of Germination*
D. aphyllum	90	–	0.0
	150	15	29.5 ± 1.57
	210	10	47.5 ± 2.01^{a}
	270	10	86.5 ± 2.79^{ab}
	330 (Dehisced)	10	56.5 ± 3.95^{abc}
	360 (Dehisced)	10	31.5 ± 2.36^{bcd}
D. chrysanthum	90	–	0.0
	150	–	0.0
	210	10	39.0 ± 2.56
	270	10	56.5 ± 4.02^{b}
	330	10	88.5 ± 2.89^{bc}
	360	10	81.5 ± 3.34^{bc}
D. chrysotoxum	90	–	0.0
	150	–	0.0
	210	15	60.0 ± 2.88
	270	15	85.5 ± 2.63^{b}
	330	15	86.5 ± 2.48^{b}
	360	15	81.0 ± 3.31^{b}
D. falconeri	90	–	0.0
	150	10	61.0 ± 2.96
	210	10	65.5 ± 2.41
	270	10	69.5 ± 4.18
	330 (Dehisced)	15	27.5 ± 2.14^{abc}
	360 (Dehisced)	15	10.5 ± 1.89^{abcd}
D. heterocarpum	90	–	0.0
	150	–	0.0
	210	20	25.5 ± 2.17
	270	15	68.0 ± 2.13^{b}
	330	15	87.5 ± 3.30^{bc}
	360	15	83.5 ± 4.15^{bc}
D. ochreatum	90	–	0.0
	150	–	0.0
	210	–	0.0
	270	20	36.0 ± 2.08
	330	20	64.0 ± 3.39^{c}
	360	20	58.0 ± 2.71^{c}

(Contd...)

Species	*Capsule Age (Days)*	*Time to Commence Germination (Days)*	*Percentage of Germination*
D. primulianum	90	–	0.0
	150	–	0.0
	210	–	0.0
	270	15	44.5 ± 2.73
	330	15	84.0 ± 3.32^{c}
	360 (Dehisced)	20	26.5 ± 2.36^{cd}
D. wardianum	90	–	0.0
	150	–	0.0
	210	15	35.5 ± 3.11
	270	10	71.5 ± 4.72^{b}
	330	10	67.0 ± 3.35^{b}
	360 (Dehisced)	10	45.0 ± 2.69^{bcd}

Values are means (n = 10)

a significantly different from 150 days

b significantly different from 210 days

c significantly different from 270 days

d significantly different from 330 days

Significance tested by students 't' test (p = 0.05)

days in case of *D. primulianum,* Depending on the species and age of the capsule seeds of various species of *Dendrobium* germinate on Kn C medium between 10-20 days of incubation. Seeds of various *Dendrobium* species obtained from 90-150 days old capsules did not germinate except in *D. aphyllum* and *D. falconeri*. In *D. ochreatum* and *D. primulianum* seed germination did not occur even in seeds obtained from 210 days old capsules. The germination of seeds occurred between 10-20 days in different species of *Dendrobium*.

In *D. aphyllum* the time taken to germinate did not differ in seeds obtained from 210-360 days old capsules. However, delayed seed germination was observed in seeds obtained from 150 days old capsules. The germination percentage increased with the capsule age upto 270 days and thereafter declined (Table 2.1, Fig. 2.1). The germination in seeds obtained from 270 days old capsules was significantly higher in comparison to seeds obtained from 150-210 and 330-360 days old capsules.

In *D. chrysanthum* the time required for germination did not differ in seeds obtained from 210-360 days old capsules. The germination percentage increased with the capsule age and maximum germination occurred in seeds obtained from 330 days old capsules. The germination of seeds declined in seeds of older cap-

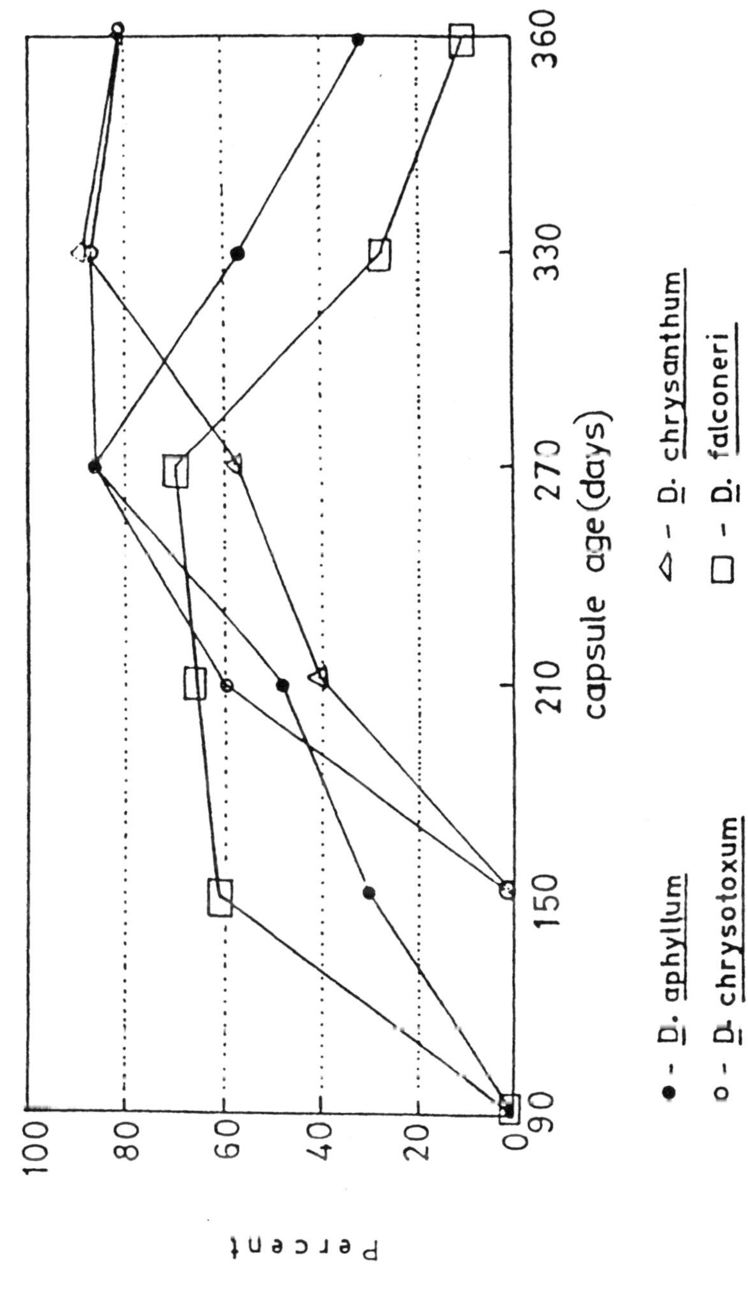

Fig. 2.1 : Capsule age and seed germination *Dendrobium*

sules. The germination of seeds obtained from 330 and 360 days old capsules was significantly higher than the germination in seeds obtained from capsules of 210 and 270 days.

In *D. chrysotoxum* the time required for germination did not differ in seeds obtained from capsules of 210-360 days old capsules. The germination percentage increased with the capsule age. Eventhough maximum germination occurred in seeds obtained from 330 days old capsules, it did not differ significantly from germination in seeds of 270 and 360 days old capsules.

In *D. falconeri* the seeds from 150-270 days old capsules germinated on the 10th day, while seeds obtained from 330-360 days old capsules germinated on the 15th day of sowing. Eventhough the germination percentage increased, with the capsule age (150-270 days) the differences in the germination percentage were not significant. Maximum seed germination occurred in seeds obtained from 270 days old capsules. The seeds of the older capsules (330-360 days) revealed a decline of the germination. The germination in seeds obtained from 330-360 days old capsules was significantly less than germination in seeds of 150-270 days old capsules.

In *D. heterocarpum* the time required for germination did not differ in seeds obtained from 270-360 days old capsules which germinated earlier than the seeds of 210 days old capsules. The germination percentage increased significantly with the capsule age. Maximum germination occurred in seeds obtained from 330 days old capsules, after which with the progression of capsule age there was a decrease in germination percentage of seeds (Table 2.1, Fig. 2.2). However, the differences in the seed germination percentage of 330 and 360 days old capsule were not significant.

In *D. ochreatum* no seed germination occurred in seeds of 90-210 days old capsules. Further, eventhough the time required for germination did not differ in seeds of 270-330 days old capsules, the germination percentage increased significantly with the capsule age. Highest germination occurred in seeds obtained from 330 days old capsules, after which there was a decrease in the germination of the seeds, with the advancement of capsule age (Table 2.1, Fig. 2.2). However, the decrease was not significantly different from the germination in seeds of 330 days old capsules.

In *D. primulianum* seeds of 90-210 days old capsules did not germinate. The seeds obtained from 270 and 330 days old capsules germinated simultaneously. However, delayed seed germination was evident in seeds obtained from 360 days old capsule. The degree of germination increased significantly with capsule age; highest germination being evident in seeds obtained from 330 days old capsules (Table 2.1, Fig. 2.2). A significant drop in the germination percentage occurred in seeds obtained from 360 days old capsules.

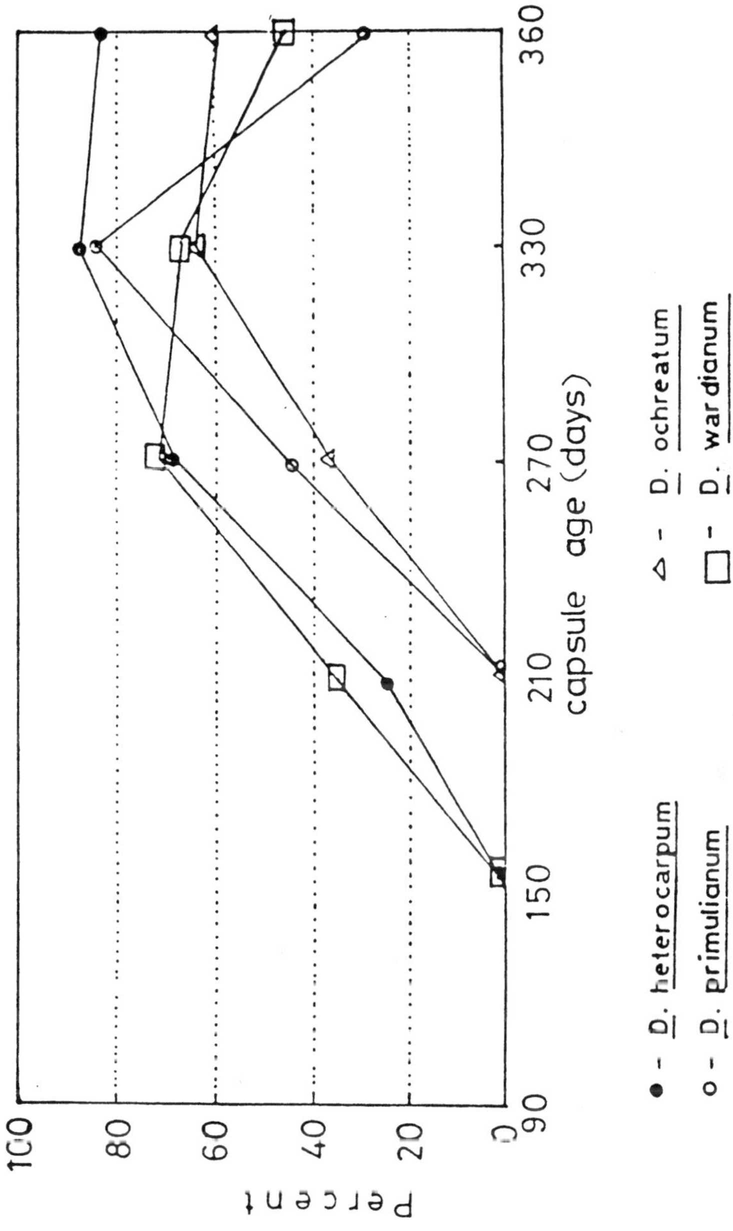

Fig. 2.2 : Capsule age and seed germination *Dendrobium*

Table 2.2 : Effect of capsule age in seed germination of *Paphiopedilum* spp. after 45 days

Species	*Capsule Age (Days)*	*Time to Commence Germination (Days)*	*Percentage of Germination*
P. fairrieanum	90	–	0.0
	150	–	0.0
	210	–	0.0
	270	30	31.5 ± 3.28
	330 (Dehisced)	35	25.5 ± 1.69^{c}
	360 (Dehisced)	35	19.0 ± 1.50^{cd}
P. hirsutissimum	90	–	0.0
	150	–	0.0
	210	–	0.0
	270	35	36.5 ± 3.17
	330	35	45.0 ± 2.69
	360 (Dehisced)	35	22.0 ± 1.69^{cd}
P. insigne	90	–	0.0
	150	–	0.0
	210	30	50.5 ± 3.20
	270	25	63.5 ± 2.79^{b}
	330	25	70.0 ± 1.97^{b}
	360	35	44.0 ± 3.14^{cd}
P. spicerianum	90	–	0.0
	150	25	53.0 ± 4.03
	210	25	78.5 ± 3.08^{a}
	270	25	73.0 ± 2.81^{a}
	330 (Dehisced)	30	52.0 ± 1.69^{bc}
	360 (Dehisced)	35	28.5 ± 1.83^{abcd}
P. venustum	90	–	0.0
	150	–	0.0
	210	35	38.0 ± 3.35
	270	35	56.5 ± 2.89^{b}
	330	35	61.5 ± 3.49^{b}
	360 (Dehisced)	35	39.5 ± 3.97^{cd}

Contd...

Species	*Capsule Age (Days)*	*Time to Commence Germination (Days)*	*Percentage of Germination*
P. villosum	90	–	0.0
	150	–	0.0
	210	35	38.5 ± 2.24
	270	35	62.0 ± 2.71^{b}
	330	35	64.5 ± 2.83^{b}
	360	35	40.5 ± 3.61^{cd}

Values are means (n = 10) ± S.E.

a significantly different from 150 days

b significantly different from 210 days

c significantly different from 270 days

d significantly different from 330 days

In *D. wardianum* seeds from 90-150 days old capsules did not germinate. The seeds obtained from capsules of 270-360 days old capsules germinated simultaneously. However, delayed seed germination was evident in seeds obtained from 210 days old capsules. The germination percentage increased significantly with the progression of age upto 270 days subsequent to which it declined (Table 2.1, Fig. 2.2). Percentage seed germination did not differ significantly in seeds of 270-330 days capsule but seeds obtained from 210 and 360 days old capsules germinated significantly less.

Paphiopedilum: Young capsules of Paphiopedilums are green which with maturity become brown. Mature capsules at any time beyond 270 days in case of *P. fairrieanum, P. spicerianum,* 330 days in case of *P. hirsutissimum, P. venustum* and 360 days in case of *P. insigne* and *P. villosum* dehisce to disperse seeds (Table 2.2).

Depending on the species and age of the capsule seeds of various species of *Paphiopedilum* germinate on Thomale GD medium 25-35 days of incubation. Seeds of various *Paphiopedilum* species obtained from capsules of 90-150 days did not germinate except in *P. spicerianum.* In all the species of *Paphiopedilum* the seeds obtained from 270 days old capsules required least time to germinate.

In *P. fairrieanum* seeds from 90-210 days old capsules did not germinate. The seeds from 270 days old capsules germinated on the 30th day, while seeds obtained from 330-360 days capsules germinated on the 35th day of sowing. Maximum germination occurred in seeds obtained from 270 days old capsules (Table 2.2, Fig. 2.3). With the further progression of capsule age seed germination diminished increasingly.

In *P. hirsutissimum* seeds obtained from 270-360 days old capsules germinated simultaneously. The germination percentage increased with the capsule age upto 330 days and thereafter declined (Table 2.2, Fig. 2.3). Maximum germination although occurred in seeds obtained from 330 days old capsule the germination of seeds did not differ from seeds of 270 days old capsules. The germination in seeds obtained from 270 and 330 days old capsules was significantly higher in comparison to seeds obtained from 360 days old capsules (Table 2.2, Fig. 2.3).

In *P. insigne* the seeds obtained from 270 and 330 days old capsules germinated earliest. Seed germination was delayed in seeds obtained from 210 and 360 days old capsules. The germination percentage increased with the capsule age upto 330 days and thereafter declined (Table 2.2, Fig. 2.3). Maximum germination occurred in seeds from 330 days old capsules. Seed germination did not differ from each other in seeds of 270 and 330 days old capsules and seeds of 210 and 360 days old capsules. However, germination of seeds in 210 and 360 days differed significantly from seed germination found in seeds of 270 and 330 days seeds.

In *P. spicerianum* seeds of 150-270 days old capsules germinated on the 25th day. The germination in seeds obtained from 330-360 days old capsules, was, however, delayed and the degree of delay was highest in 360 days old capsules. The germination percentage increased with the capsule age and maximum germination occurred in seeds of 210 days old capsules (Table 2.2, Fig. 2.3). The germination in seeds of 210 and 270 days old capsule did not differ from each other. The germination of seeds in seeds of 150, 330 and 360 days old capsules, however differed significantly from germination in seeds of 210 days old capsules.

In *P. venustum* seeds obtained from 210-360 days old capsules germinated simultaneously. Seeds of 90 days old capsules did not germinate. The germination percentage increased with the capsule age (Table 2.2, Fig. 2.4). Maximum germination occurred in seeds obtained from 330 days old capsules (Table 2.2, Fig. 2.4). Germination in seeds of 270 and 330 days old capsules did not differ from each other. But germination in seeds obtained from 210 and 360 days old capsules differed significantly from germination in seeds of 330 days old capsule.

In *P. villosum* seeds obtained from 210-360 days old capsules germinated simultaneously. The germination of seeds increased with the capsule age. Maximum germination occurred in seeds obtained from 330 days old capsules (Table 2.2, Fig. 2.4). But the germination in seeds of 270 and 330 days old capsule did not differ from each other. However, the germination in seeds of 210 and 360 days old capsules differed significantly from seed germination in seeds of 330 days old capsules.

DISCUSSION

Seeds of Dendrobiums germinated earlier in comparison to Paphiopedilums.

Within the genus *Dendrobium*, germination occurred earliest in *D. aphyllum*,

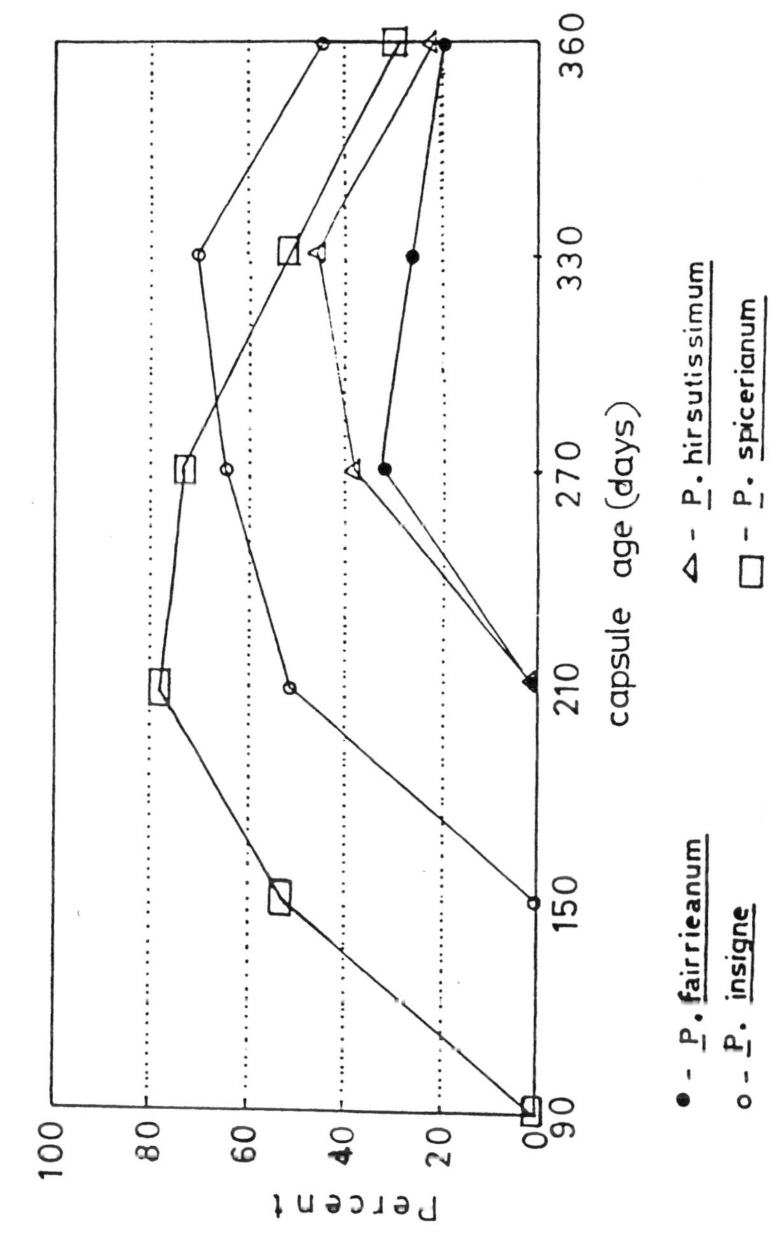

Fig. 2.3 : Capsule age and seed germination *Paphiopedilum*

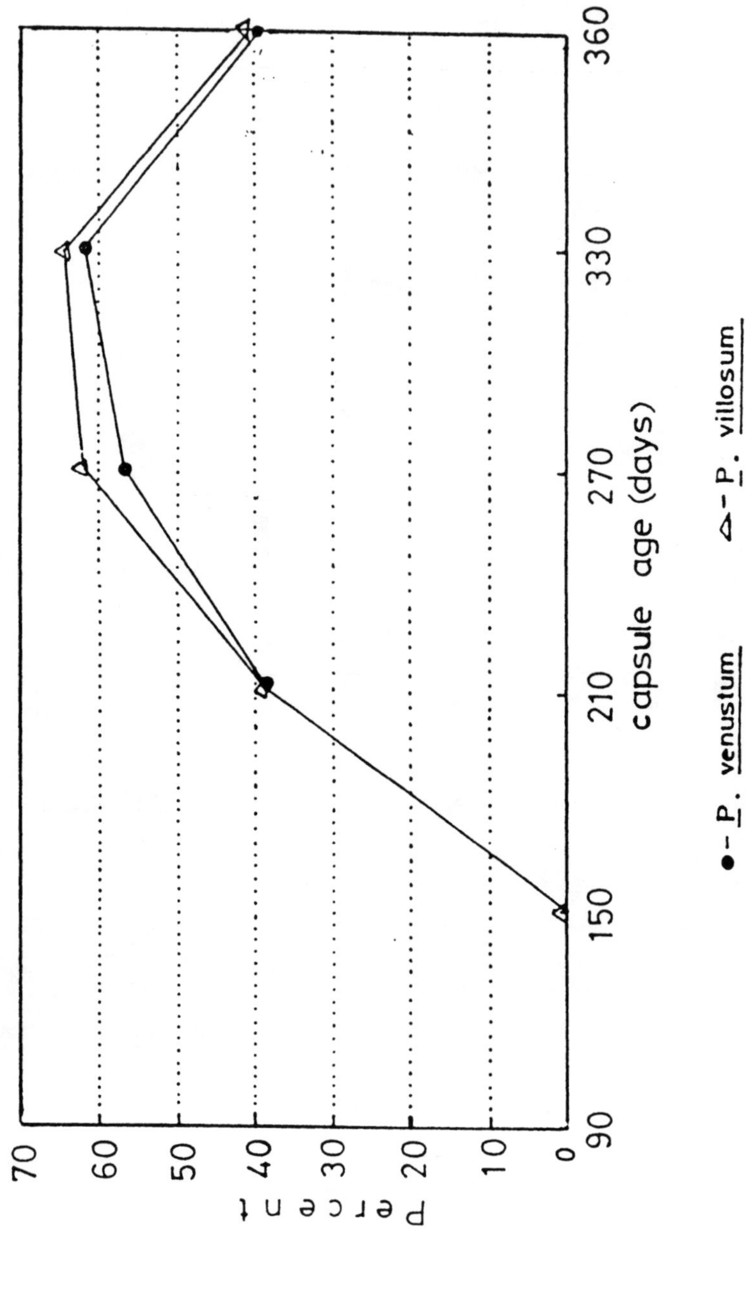

Fig. 2.4 : Capsule age and seed germination *Paphiopedilum*

D. chrysanthum, D. falconeri and *D. wardianum, D. ochreatum* seeds germinated last. The seeds of *D. chrysotoxum, D. heterocarpum* and *D. primulianum* were intermediate in germinating. Delay in commencement of germination was recorded in seeds from young capsules of *D. aphyllum, D. heterocarpum* and *D. wardianum*. This delay could be attributed to the fact that the embryo in the younger capsules had not sufficiently advanced in development for germination. Similar findings were reported by Ballard (1987) for *Cypripedium reginae*. The critical stage of development of embryo required for germination is species dependent. In all the species of *Dendrobium* investigated seeds from 90 days old capsules never germinated. Seeds from 150 days old capsules also did not germinate except in *D. aphyllum* and *D. falconeri*. The seeds obtained from 210 days old capsules in most of the species germinated (*D. chrysanthum, D. chrysotoxum, D. heterocarpum* and *D. wardianum*) except *D. ochreatum* and *D. primulianum* where seed germination occurred only after the capsules were 270 days old.

In *D. aphyllum, D. falconeri, D. heterocarpum* and *D. wardianum* maximum germination occurred in seeds of 270 days old capsules but in *D. chrysanthum, D. chrysotoxum, D. ochreatum, D. primulianum*, the maximum germination of seeds occurred in seeds obtained from 330 days old capsules. The decrease in germination percentage of seeds after reaching an optima could be due to the onset of dormancy in seeds before the dehiscence of capsules. This is also reported, for *Aerides multiflorum, Rhynchostylis retusa, Vanda parviflora* (Vij *et al.*, 1981), *Cypripedium reginae* (Curtis, 1943; Withner, 1953; Stoutamire, 1974 and Ballard, 1987) and *Dactylorchis* (Vermeulen, 1947).

Within the genus, *Paphiopedilum*, the seeds of *P. insigne, P. spicerianum* were first to germinate and seeds of *P. hirsutissimum, P. venustum* and *P. villosum* germinated last. The seeds of *P. fairrieanum* were intermediate in germination. Within the species germination varied with the age of the capsule. In all the species no germination occurred in seeds of 90 days old capsules. Except for *P. spicerianum*, in all the other species of *Paphiopedilum*, the seeds obtained from 150 days old capsules also did not germinate. Seeds obtained from 210 days old capsules did not germinate in *P. fairrieanum* and *P. hirsutissimum*. These findings suggest that developing embryo in different species reaches the critical stage of development necessary for seed germination at different times. In Paphiopedilums also seed germination gradually increased with increasing age of the capsule upto a threshold age and then decreased. Assimilation of inhibitors in seed coat (Burgeff, 1954) in addition to onset of dormancy (Vij *et al.*, 1981) could be attributed for the decrease in germination of seeds from capsules of Paphiopedilums. Thus, the present findings suggest that capsule age is very important factor for achieving optimum seed germination and it is species specific. The investigation thus supports the conclusions of Arditti *et al.*, (1982) which suggests that the earliest stage at which embryos can be cultured successfully varies with genus, species, hybrids and local conditions.

3
Effect of Nutrient Medium on Seed Germination and Seedling Growth

Introduction

Orchid seeds are produced in large numbers within a capsule. The seeds are very minute and contain only a tiny sphere of undifferentiated cells possessing little or no endosperm (nutritive) tissue. However, some stored energy is present in the embryo in the form of fat droplets and starch granules (Burgeff, 1936; Ernst *et al.,* 1970). Due to the non-endospermic nature of the orchid seed, their germination in nature is a unique phenomenon.

Bernard (1904) established that seeds germinated successfully only when infected by a suitable fungi. Subsequently, Knudson (1922, 1924, 1925) achieved asymbiotic germination of the seeds on a simple medium containing sucrose which led to the conclusion that fungus was responsible for breaking the complex starch into readily utilisable simple carbohydrates for the orchid seed germination. Knudson (1930) raised *Cattleya* plants *in vitro* from seeds, and maintained these till blossom (6-7 years) thus demonstrating that an obligate symbiotic relationship with the fungus was not a must for the germination of orchid seeds. Since then a number of orchid seeds have been successfully germinated using different nutrient media (Arditti, 1967; Arditti *et al.* 1982; Stoutamire, 1981; Arditti & Ernst, 1984; Yam and Weatherhead, 1988).

The nutritional requirement varies with the genus, species and locality (Arditti *et al.* 1982; Arditti & Ernst, 1984). Many media differing in chemical composition have been developed for general and specific uses, e.g., Burgeff N_3F (Burgeff, 1936), Fast (Fast, 1976), Knudson C (Knudson, 1946), Murashige and Skoog (Murashige and Skoog, 1962), Pfeffer (Harvais, 1972), Robert Ernst (Ernst, 1982), Thomale GD (Thomale, 1954), Vacin and Went (1949), Wolter and Skoog (Wolter, 1968), Zak (Borriss, 1969). Ionic concentration of the medium also affects germination (Arditti *et al.*, 1982).

In the present study, an effort was made to find out which one of the aforementioned medium is most suitable for the germination of seeds and induction of shoots and roots in *D. chrysanthum* and *P. spicerianum.*

MATERIALS AND METHODS

Capsules of *Dendrobium chrysanthum* (330 days old) and *Paphiopedilum spicerianum* (270 days old) were collected. The capsules were surface sterilized by immersing them in ethyl alcohol (90%) for a few seconds and flaming. The surface sterilized capsules were cut open with sterilized scalpel blade to take out seeds for germination *in vitro* on different media. Approximately fifty seeds were sown in a tube containing 20 ml medium. Ten tubes were used for each medium.

Table 3.1 : Composition of some media (mg 1^{-1}).

Components	*N_3F*	*F*	*KnC*	*MS*	*RE*	*Th GD*	*VW*	*WS*	*Pf*	*Z*
MACRO-ELEMENT										
$(NH_4)_2 SO_4$	250	–	500	–	150	60	500	–	–	–
$NH_4 NO_3$	–	160	–	1650	400	370	–	–	–	–
$NaNO_3$	–	–	–	–	–	–	–	4800	–	–
KNO_3	–	–	–	1900	400	400	525	70	200	–
$Ca(NO_3)_2 . 4H_2O$	1000	80	1000	–	150	–	–	613	800	–
$Mg(NO_3)_2 . 6H_2O$	–	–	–	–	100	110	–	–	–	–
$Ca_3 (PO_4)_2$	–	–	–	–	–	–	200	–	–	–
$MgSO_4 . 7H_2O$	250	80	250	370	–	–	250	1550	200	–
$KH_2 PO_4$	–	80	250	170	300	300	250	–	200	–
$K_2 HPO_4$	250	–	–	–	–	–	–	–	–	–
$CaCl_2 . 2H_2O$	–	–	–	440	–	–	–	–	·	–
KCl	250	160	–	–	–	–	–	142	100	–
$NaH_2PO_4 . H_2O$	–	–	–	–	–	–	–	400	–	–
$Na_2 SO_4 . 10 H_2O$	–	–	–	–	–	–	–	1388	–	–
MICRO-ELEMENT										
$FeSO_4 . 7H_2O$	20	20	25	27.8	25	20	–	–	25	–
$Fe_2 (C_2H_4O_6)_3$	–	–	–	–	–	–	28	28	–	–
$MnSO_4 . 4H_2O$	–	0.1	7.5	22.3	–	–	7.5	0.5	0.5	–
$H_3 BO_3$	–	1.0	0.056	6.3	–	–	–	3.2	0.5	–
$ZnSO_4 . 7H_2O$	–	1.0	0.331	8.6	–	–	–	3.2	0.5	–
$CuSO_4 . 5H_2O$	–	0.03	0.040	0.025	–	–	–	–	0.5	–
$AlCl_3$	–	0.03	–	–	–	–	–	–	–	–
Na_2 EDTA	–	–	–	37.3	–	–	–	5.5	–	–
$ZnCl_2$	–	–	–	–	–	–	–	–	–	–
KI	–	0.01	–	0.83	–	–	–	1.6	–	–
$CoCl_2$	–	–	–	0.025	–	–	–	–	–	–
$Na_2 MoO_4 . 2H_2O$	–	–	–	0.025	–	–	–	–	–	–
$NiCl_2 . 6H_2O$	–	0.03	–	–	–	–	–	–	–	–
VITAMINS										
Thiamine HCl	–	–	–	–	–	–	–	0.1	·	–
Myo-inositol	–	–	–	100	–	–		100		
COMPLEX ADDITIVES										
Citric acid	90	–	–	–	–	–	–	–	–	–
Peptone	–	1.5	–	–	–	–	–	–	–	–
Fresh yeast	–	8.3	–	–	–	–	–	–	–	–
Coconut milk	–	–	–	–	–	–	–	–	–	20 ml
SUGARS										
Glucose	10000	–	–	–	–	–	–	10000	10000	–
Fructose	10000	5000	–	–	20000	–	–	–	–	–
Sucrose	–	12000	20000	30000	–	20000	20000	–	–	10000
p H	5.3	5.5	5.0	5.5	5.2	5.1	5.2	5.5	5.0	5.0
AGAR	12000	8000	12000	8000	16000	13000	12000	8000	10000	12000

Chemical compositions and pH of the various nutrient media N_3F, Burgeff N_3F medium (Burgeff, 1936); F, Fast medium (Fast, 1976); Kn C, Knudson C medium (Knudson, 1946); MS, Murashige and Skoog medium (Murashige and Skoog, 1962); RE, Robert Ernst medium (Ernst, 1982); Th GD, Thomale GD medium (Thomale, 1954); VW, Vacin and Went medium (Vacin and Went, 1949); WS, Wolter and Skoog medium (Wolter, 1968); Pf, Pfeffer medium (Harvais, 1972) and Z, Zak medium (Borriss, 1969), used for germinating seeds are given in Table 3.1. Twenty ml medium was poured into 25 × 200 mm Borosil culture tubes. The media were sterilized by autoclaving for 20 minutes at 121°C and 15 lb/in^2 pressure. The cultures of *D. chrysanthum* were incubated in continuous light (3,000 lux) at 25 ± 1°C while the cultures of *P. spicerianum* were initially kept in continuous dark but five days after commencement of germination the cultures were kept in continuous light (3,000 lux) at 25 ± 1°C for greening of protocorms and subsequent seedling development. Time taken for commencement of germination by seeds and time required for the greening of *P. spicerianum* protocorms was recorded at 5 days interval.

Detailed Method of Sampling and Recording of Data

Germination percentage of seeds for *D. chrysanthum* was recorded on the 30th day and for *P. spicerianum* on the 50th day. For finding out germination percentage the germinated and ungerminated seeds from the five tubes of each individual treatment were scooped out and placed in a petriplate and mixed together. Ten slides were prepared by placing the germinated and ungerminated seeds in a drop of water and covering it with a cover slip. Per slide one microscopic field (approximately 20 seeds) was scrutinized randomly under a stereoscopic binocular microscope, for recording the germination percentage. Altogether, 200 seeds were scrutinized per treatment.

To study seedling growth the differentiated protocorms were scooped out from the remaining five tubes of respective medium 120 days after culturing and placed on a petriplate. Ten most developed seedlings of each treatment were selected for recording data on the various growth parameters (shoot height, leaf number, leaf length, leaf breadth, root number, root length and fresh weight of the differentiated protocorms).

RESULTS

Seeds in these experiments, germinated *in vitro* (Plate 3a. 1-11) by forming protocorms (Plate 3a. 1-4, 7-9). During germination, in both the taxa, the embryo within the seed coat increased in size (Plate 3a.2 and 8). This was followed by the splitting of the seed coat and emergence of the protocorms (Plate 3a. 3, 4).

The effect of different medium on seed germination and seedling growth of *D. chrysanthum* (Plate 3b. 1-9) and *P. spicerianum* (Plate 3b.10-18) are summarized in Table 3.2 & 3.3.

Table 3.2 : Effect of various nutrient medium on seed germination and seedling growth in *D. chrysanthum.*

Character	*Medium*									
	N_3F	*F*	*KnC*	*MS*	*Pf*	*RE*	*ThGD*	*VW*	*WS*	*Z*
Time taken to commence germination (days)	10	10	10	10	10	10	10	10	-	25
Germination Percent	88.5± 3.1	[a] 73.0± 3.5	[b] 87.0± 3.2	[abc] 44.5± 2.7	[abc] 45.0± 2.5	[abcde] 66.5± 3.2	[bdef] 92.0± 2.9	[acdeg] 69.5± 4.2	-	[abcdefghi] 29.5± 1.6
Shoot Height (mm)	5.1± 0.4	-	[a] 8.4± 0.4	-	-	[ac] 1.4± 0.1	[acf] 6.2± 0.2	-	-	-
Leaf:										
Number	1.2± 0.1	-	[a] 2.0± 0.0	-	-	[ac] 1.0± 0.0	[af] 1.8± 0.1	-	-	-
Length (mm)	2.4± 0.2		[a] 4.5± 0.3	-	-	[ac] 0.9± 0.1	[af] 4.5± 0.2	-	-	-
Root:										
Number	-	-	1.7± 0.1	-	-	-	[c] 0.3± 0.1	-	-	-
Length (mm)	-	-	3.2± 0.2	-	-	-	[c] 0.3± 0.1	-	-	-
Seedling fresh weight (mg)	5.1± 0.3	-	[a] 8.8± 0.6	-	-	[ac] 1.4± 0.1	[acf] 6.8± 0.2	-	-	-

Values are means (n = 10) ± S.E.

a significantly different from N_3F
b significantly different from F
c significantly different from KnC
d significantly different from MS
e significantly different from Pf
f significantly different from RE
g significantly different from Th GD
h significantly different from VW
i significantly different from WS

Significance tested by students 't' test (p = 0.05)

D. chrysanthum: D. chrysanthum seeds germinated into green protocorms on all media except WS. The germination of seeds commenced on 10th day on all media except Z, where seeds germinated on the 25th day. The germination of the *in vitro* cultured seeds, on the various media, decreased in the following order : Th GD > N_3F > Kn C > F > VW > RE > Pf > MS > Z (Fig. 3.1) but the germination of seeds did not differ from each other in the seeds germinated on Th GD, N_3F and Kn C media. The protocorms formed by seeds germinated on the F, MS, Pf, VW and Z media did not differentiate shoots and roots (Plate 3b. 5-9). However, protocorms of seeds germinated on N_3F, Kn C, RE and Th GD medium developed shoots (Plate 3b. 1-4). The shoot height significantly differed from each

Plate 3a : Stages during germination *in vitro* of *P. spicerianum* (Fig. 1-6) and *D. chrysathum* (Fig. 7-11)

1. seeds at the time of inoculation ×100
2. enlarged embryo within the seed coat× 100
3. splitting of seed coat and emergence of protocorm × 100
4. protocorm × 100
5. protocorm with vegetative apex × 100
6. seedling with differentiated leaf × 50
7. seed at the time of inoculation × 400
8. enlarged embryo with in seed coat × 400
9. splitting of seed coat and emergence of protocorm × 150

10-11. differentiation of the vegetative apex × 100

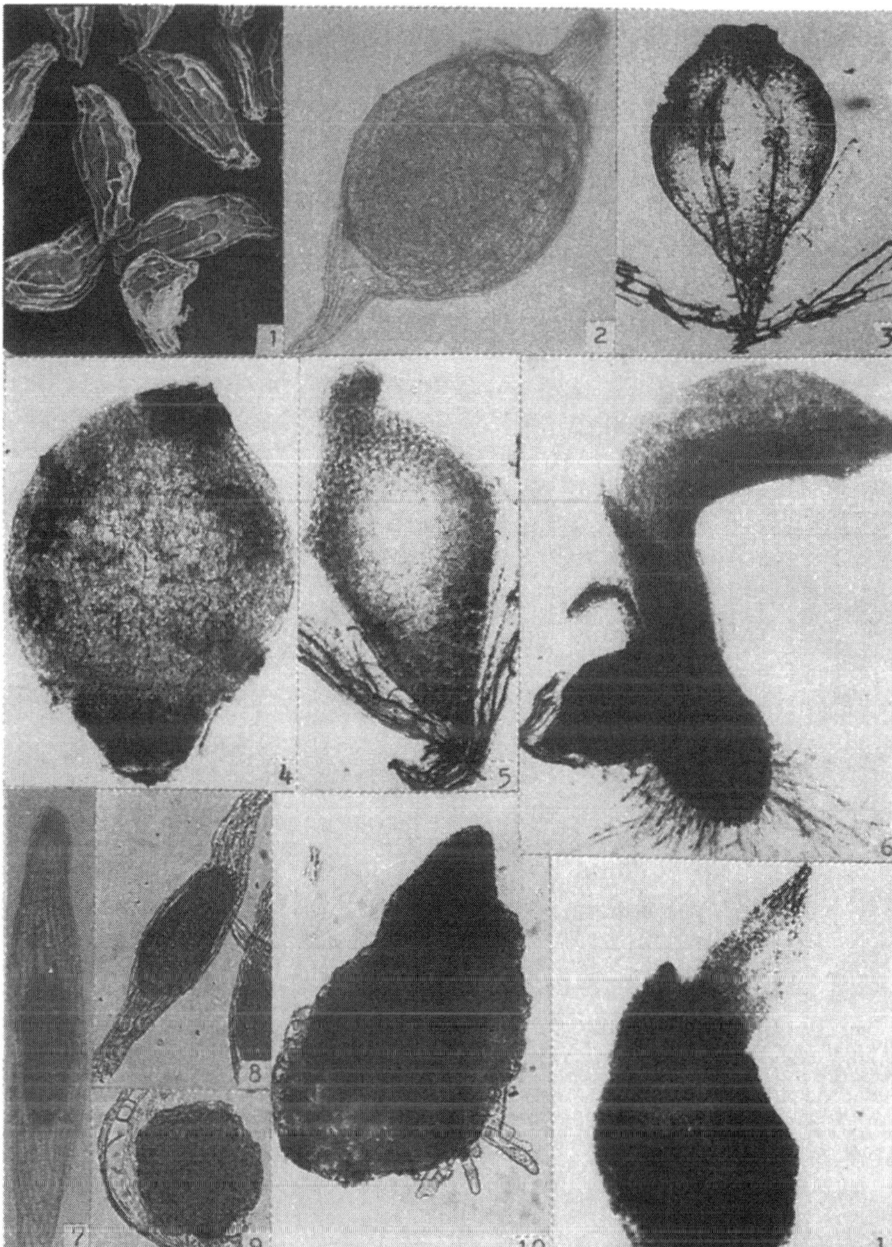

Plate 3b : Effect of various nutrient medium on seed germination and seedling growth of

D. chrysanthum (Fig. 1-9) and *P. spicerianum* (Fig. 10-18)

1. N_3F medium
2. Th GD medium
3. Kn C medium
4. RE medium
5. F medium
6. VW medium
7. Pf medium
8. MS medium
9. Z medium
10. Th GD medium
11. N_3F medium
12. F medium
13. Z medium
14. MS medium
15. VW medium
16. RE medium
17. Kn C medium
18. Pf medium

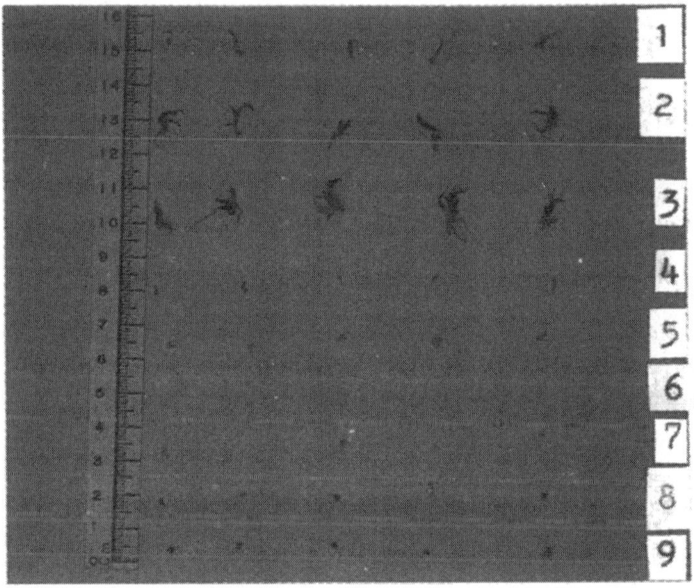
1
2
3
4
5
6
7
8
9

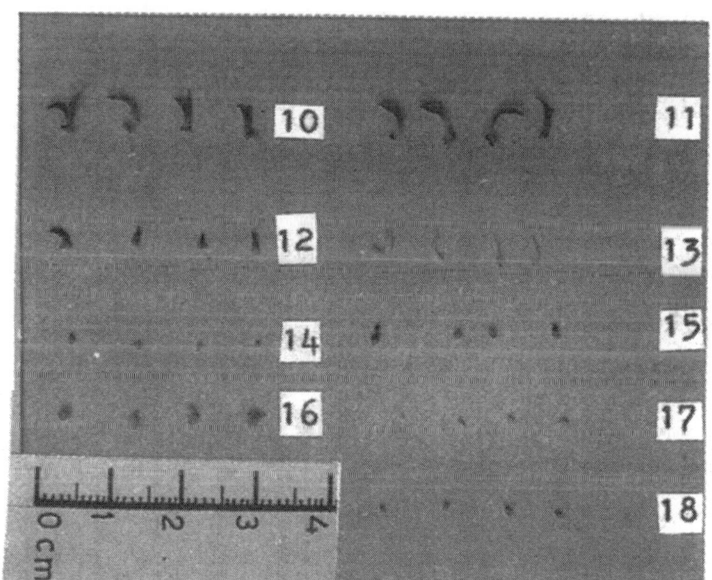
10
11
12
13
14
15
16
17
18
0 cm 1 2 3 4

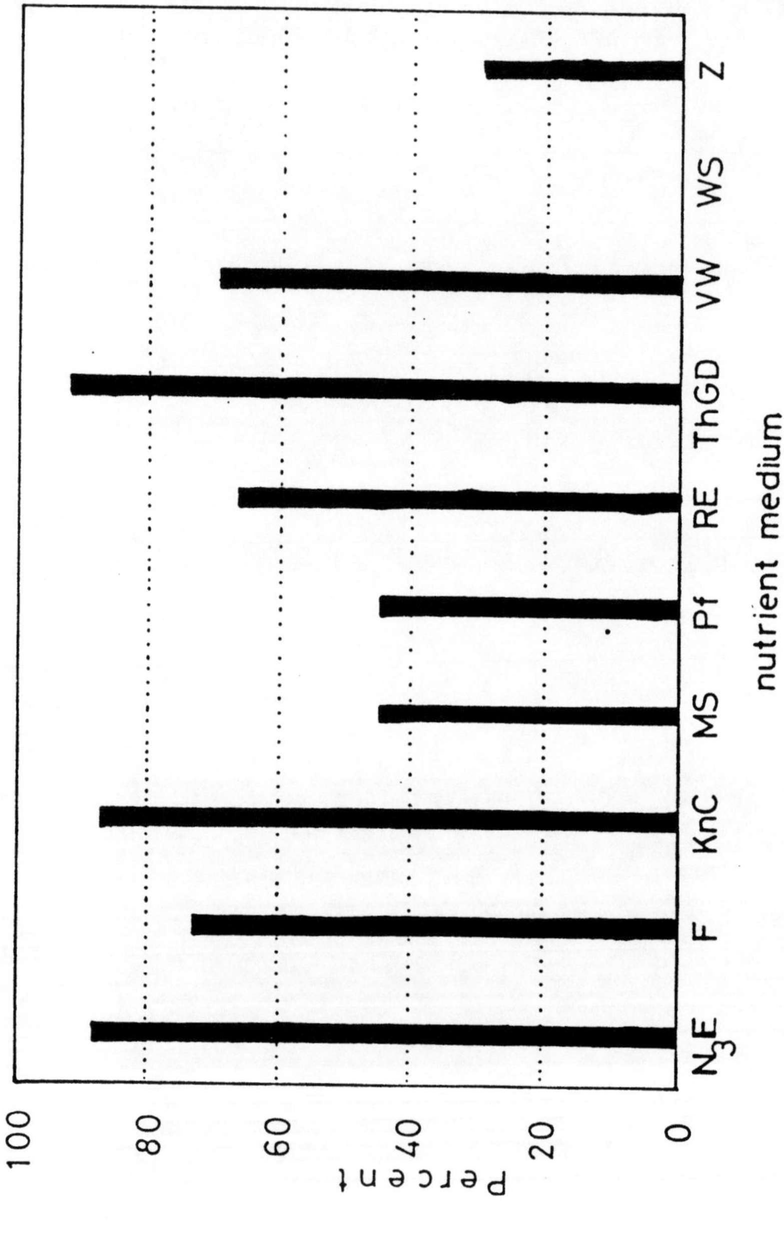

Fig. 3.1: Nutrient medium and seed germination *Dendrobium*

other and the order of decrease in shoot height was Kn C > Th GD > N_3F > RE (Table 3.2). Maximum number of leaves developed on Kn C medium but it did not differ significantly from the leaves developed on Th GD medium (Table 3.2).

Table 3.3 : Effect of various nutrient medium on seed germination and seedlng growth in *P. spicerianum.*

Character	*Medium*									
	N_3F	*F*	*KnC*	*MS*	*Pf*	*RE*	*ThGD*	*VW*	*WS*	*Z*
Time taken to commence germination (days)	30	45	45	40	40	30	25	30	–	30
			a	abc	abcd	abc	bcdef	cdefg		acdefg
Germination per cent	61.5± 2.9	58.0± 2.1	4.0± 3.9	20.5± 2.0	12.5± 2.0	17.5± 2.3	67.5± 2.8	59.0± 2.2	– –	52.5± 2.5
Time taken for the greening of protocorms	10	10	–	–	–	5	5	10	–	10
		a				a	bf	abg		bfh
Shoot height (mm)	4.9± 0.0	2.4± 0.1	–	–	–	2.9± 0.1	5.0± 0.2	2.9± 0.1	–	5.2± 0.6
Leaf:										
Number	1.1± 0.1	1.0± 0.0	–	–	–	2.0± 0.0	2.0± 0.0	1.0± 0.0	–	1.3± 0.0
		a				a	bf	abfg		abfg
Length (mm)	4.1± 0.0	0.8± 0.0	–	–	–	0.8± 0.0	3.9± 0.3	0.3± 0.1	–	0.9± 0.0
Root:										
Number	–	–	–	–	–	–	1.0± 0.0	–	–	–
Length (mm)	–	–	–	–	–	–	0.4± 0.0	–	–	–
						ab	abf	abfg		abfgh
Seedling fresh weight (mg)	1.1± 0.0	1.0± 0.1	–	–	–	1.3± 0.0	2.9± 0.1	1.7± 0.1	–	0.5± 0.0

Values are means (n = 10) ± S.E.

a significantly different from N_3F
b significantly different from F
c significantly different from KnC
d significantly different from MS
e significantly different from Pf
f significantly different from RE
g significantly different from ThGD
h significantly different from VW
i significantly different from WS

Significance tested by students 't' test (p = 0.05)

The N_3F and RE media, however, significantly decreased leaf development (Table 3.2). Longest leaves developed on Kn C and Th GD media. A significant decrease in leaf length was recorded on N_3F and RE media (Table 3.2). Root differentiation was recorded only on Kn C and Th GD media (Plate 3b. 2, 3b. 3). Maximum number of roots and longest roots developed on Kn C medium (Table 3.2). The fresh weight of seedling developing on Kn C medium had maximum fresh weight

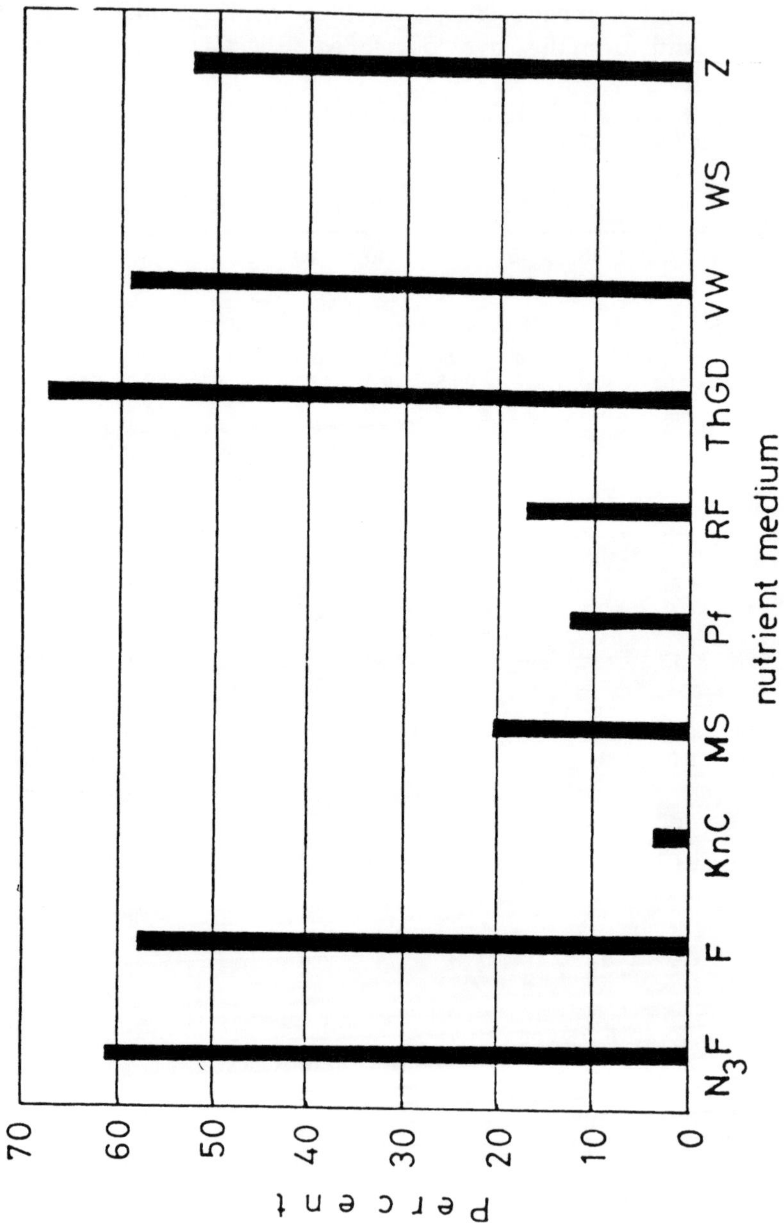

Fig. 3.2 : Nutrient medium and seed germination *Paphiopedilum*

which significantly differed from the fresh weight of seedlings grown on Th GD medium. The fresh weight of the seedling differentiated on the N_3F and RE also differed significantly from each other (Table 3.2); the seedlings grown on N_3F medium being heavier. However, their fresh weight was significantly less in comparison to seedlings grown on Kn C and Th GD media (Table 3.2).

P. spicerianum: P. spicerianum seeds germinated *in vitro* developed white protocorms on all media except WS medium. Commencement of seed germination occurred earliest (25th day) on the Th GD medium. On the 30th day, seeds germinated on N_3F, RE, VW and Z media; on 40th day on MS and Pf medium and on 45th day on F and Kn C medium (Table 3.3). Percentage of germination on various medium was in the following order: Th GD > N_3F > VW > F > Z > MS > RE > Pf > Kn C (Fig. 3.2). The germination of seeds was significantly higher on Th GD medium compared to other media except N_3F medium (Table 3.3). Time taken by the protocorms for greening after being kept in light was minimum on RE and the Th GD media (5 days) as against 10 days required for greening of protocorms on N_3F, F, VW and Z media; the protocorms on Kn C, MS and Pf media remained white (Table 3.3).

Shoot elongation was maximum on Z medium, but it did not differ significantly from the shoot height recorded on Th GD and N_3F media. The shoot height decreased in the following order: Z > Th GD > N_3F > RE > VW > F (Table 3.3).

Table 3.4 : Major element composition (in millimoles) of several media used for orchid germination and seedling growth.

IONS	*N_3F*	*F*	*KnC*	*MS*	*Pf*	*RE*	*ThGD*	*VW*	*WS*	*Z*
Nitrate	8.40	2.77	8.40	39.40	8.73	11.57	9.45	5.19	66.85	–
Ammonium	3.80	2.07	7.60	20.61	–	7.27	5.50	7.56	59.99	–
Nitrate: Ammonium ratio	2.20	1.34	1.10	1.90	–	1.59	1.82	0.69	1.11	–
Phosphate	1.40	0.01	1.00	1.24	1.46	2.20	2.20	3.14	2.89	–
Sulphate	2.90	0.34	4.80	1.50	0.81	1.22	2.16	4.83	5.87	–
Chloride	3.40	2.22	–	5.98	1.34	–	–	–	1.90	–
Potassium	6..20	2.83	1.80	20.03	4.77	6.16	6.16	7.03	3.58	–
Magnesium	1.00	0.34	1.00	1.50	0.81	0.67	0.74	1.01	5.87	–
Calcium	4.20	0.35	4.20	2.99	3.38	0.63	–	1.95	2.59	–
Citrate	0.48	–	–	–	–	–	–	–	–	–
Iron	0.07	0.01	0.09	–	–	0.09	0.07	0.19	–	–
Manganoce	–	–	0.03	–	–	0.03	–	0.04	0.05	–
Sodium	–	–	–	–	–	–	–	–	10.63	–
Total concentrations	28.17	11.54	29.72	93.2	21.30	30.46	24.30	30.94	166.24	0.00
Ranking	(5)	(2)	(6)	(9)	(3)	(7)	(4)	(8)	(10)	(1)

Maximum number of leaves developed in seedlings grown on Th GD and RE media. The number of leaves however decreased in the following order: Th GD = RE > Z > N_3F = F = VW (Table 3.3). Leaf length was maximum on N_3F medium, followed by Th GD medium. Leaves were significantly smaller on other media (Table 3.3).

Roots differentiated only on Th GD medium (Table 3.3). Fresh weight of the seedlings was maximum on Th GD medium, but it did not differ significantly from fresh weight of seedlings developing on N_3F medium (Table 3.3).

Table 3.4 which shows the ion concentrations of the various media used, reveals that both the taxa can germinate on a wide variety of inorganic salt combinations and concentrations i.e.; 11.54 to 93.25 millimoles. However, the two taxa preferred two different medium for better seed germination and seedling growth. Knudson C medium, found better for *D. chrysanthum,* had a higher ion concentration than Thomale GD medium found better for *P. spicerianum* (Table 3.4).

DISCUSSION

During the present investigation, seeds of both the taxa germinated and developed seedlings on various media used. However, Knudson C medium having inorganic ion concentration 29.72 millimoles proved better for the germination of seeds and growth of *D. chrysanthum* seedlings. Arditti and Ernst, (1984) also reported that orchid seeds germinate on a wide variety of inorganic salt combinations and ion concentrations, but in general a balanced mineral mixture and medium concentration is an important factor for every orchid species. Better seedling growth of *Dendrobium nobile* on Knudson C medium has also been reported by Raghuwanshi *et al.,* (1985). Thomale GD medium having inorganic ion concentration 24.3 millimoles proved better for the terrestrial species *P. spicerianum*. Suitability of Thomale GD medium for *Paphiopedilum* seed germination has also been reported by Lugo (1955) and Flamee (1978). Thomale GD medium for seedling growth has also been used as basal medium by Lugo (1955) and Flamee (1978). However, Ernst (1982) reported successful seed germination and seedling growth on RE medium. While in the present investigation seed germination and seedling growth on RE medium having inorganic ion concentration 30.46 millimoles was found to be very poor. These findings thus suggest that, in comparison to *P. spicerianum, D. chrysanthum* prefers a medium having higher ion concentration. Harbeck (1963), Harvais (1973), Fast (1976), V th (1976), Arditti and Ernst (1984) have reported that for terrestrial orchids, especially the lady's slipper (*Cypripedium, Paphiopedilum*) medium with low ion concentration i.e., a more dilute medium is better. The present investigation supports these conclusions.

4
Effect of Light and pH on Seed Germination and Seedling Growth

Introduction

Light affects seed germination in many plants, e.g., orchids (Arditti, 1967, 1979; Arditti and Ernst, 1984). Epiphytic orchid species although may germinate in both light and dark, they sometimes require light for induction and improvement of shoot and/or root (Arditti and Ernst, 1984). But in terrestrial species such as *Cypripedium calceolus* (Fast, 1976); *C. reginae, C. irapeanum, C. californicum, C. acaule, C. candidum, Coeloglossum viride, Gymnadenia* spp., *Orchis militaris, Dactylorhiza* spp., *Barlia longibracteata, Platanthera hyperborea* and *Corycium orobanchiodes* (Stoutamire, 1974); *Serapis parviflora* and *S. orientalis* (Vöth, 1976), and *Paphiopedilum callosum* (Kano, 1965) light causes inhibition.

The need of an appropriate pH for germination of orchid seeds was emphasised by Knudson (1924) and Clement (1924) and Goh (1971). Even though the orchid seeds germinate between pH 3.6-7.6, the optimum pH for orchid seed germination is reported to be 4.8 to 5.2 (Arditti, 1967, 1979; Arditti and Ernst, 1984). pH is critical only during the early stages of germination (Arditti, 1967, 1979; Withner, 1951) because seedlings are less sensitive to variation in pH (Knudson, 1951; Ernst, 1967, 1974; Miyazaki and Nagamatsu, 1965).

The present investigation deals with the effect of light and pH on seed germination and seedling growth in *D. chrysanthum*, an epiphytic species and *P. spicerianum*, a terrestrial species.

MATERIALS AND METHODS

Capsules of *D. chrysanthum* (330 days old) and *P. spicerianum* (270 days old) were collected. The capsules were surface sterilized by immersing in ethyl alcohol (90%) for a few seconds and flaming. The seeds from surface sterilized capsules were taken out by cutting open the capsules with sterilized scalpel. The seeds of *D. chrysanthum* were germinated on Knudson C medium and seeds of *P. spicerianum* were germinated on Thomale GD medium. Preparation of the two medium is described in chapter 3. Approximately fifty seeds were sown per culture tube containing 20 ml medium. Per treatment ten tubes were used. To find out requirement of light for *D. chrysanthum* seed germination ten tubes were kept in continuous light and ten tubes in continuous dark. In case of *P. spicerianum* ten tubes were kept in continuous light, ten tubes in continuous dark and ten tubes were first kept in dark and after five days from the commencement of germination these were placed in light. For the pH studies, pH of the two medium (Knudson C medium and Thomale GD medium) were variously adjusted

(3.0, 4.0, 4.5, 5.0, 5.5, 6.0 and 6.5) by adding 1N HCl and 1N NaOH solution. Preparation and sterilization of medium is described in chapter 3.

Ten tubes per treatment were maintained. For pH studies, cultures of *D. chrysanthum* were kept under continuous light (2,500-3,000 lux) and that of *P. spicerianum* in dark; five days after the commencement of germination these were transferred to light. The cultures were maintained at 25 ± 1°C temp., and observations were recorded upto 90 days.

Detailed Method for Recording of Data

For Light Requirement Studies: For both the taxa the following parameters were considered; (a) time taken to commence germination, (b) time taken for greening of protocorms, (c) gemination percentage of seed (methods for sampling and recording of data described in chapter 3), (d) growth assessments were made on the 90th day of *in vitro* culture of seed. Characters taken for growth assessment were: protocorm stage (+), seedlings with differentiated leaf (++), seedlings exhibiting leaf and root differentiation (+++). The colour of the seedling and survival percentage was also recorded on the 90th day. For recording per cent seedling survival, seedlings from five tubes of each treatment were mixed together and per treatment 200 seedlings were scrutinized randomly.

Data Recording for pH Studies: For both the taxa the following parameters were considered; (a) time taken to commence germination, (b) germination percentage of seed (methods for sampling and recording of data described in chapter 3), (c) growth assessments were made on the 90th day of *in vitro* culture of seed. Characters taken for growth assessment were: protocorm stage (+), seedlings with differentiated leaf (++), seedlings exhibiting leaf and root differentiation (+++), and (d) colour of the seedlings was also recorded on the 90th day.

Table 4.1 : Effect of light on seed germination, seedling growth, pigmentation and survival of the seedlings in *D. chrysanthum* and *P. spicerianum*.

Character	*D. chrysanthum*		*P. spicerianum*		
	Light	*Dark*	*Light*	*Dark*	*Dark for 25 days then kept in light*
Time taken to commence germination (days)	10	10	–	25	25
Germination per cent	87.5	80.5	–	71.0	71.0
Time taken for the greening of protocorms (days)	Germinated protocorms were green	Remained white		Remained white	5
Differentiation after 90 days	+++	++	–	++	+++
Pigmentation of seedling	MG	W	–	W	MG
Per cent seedling survival after 90 days	81.5	0	–	0	66.5

+ protocorm stage
++ seedlings exhibiting leaf differentiation
+++ seedlings exhibiting leaf and root differentiation
MG moderately green
W white

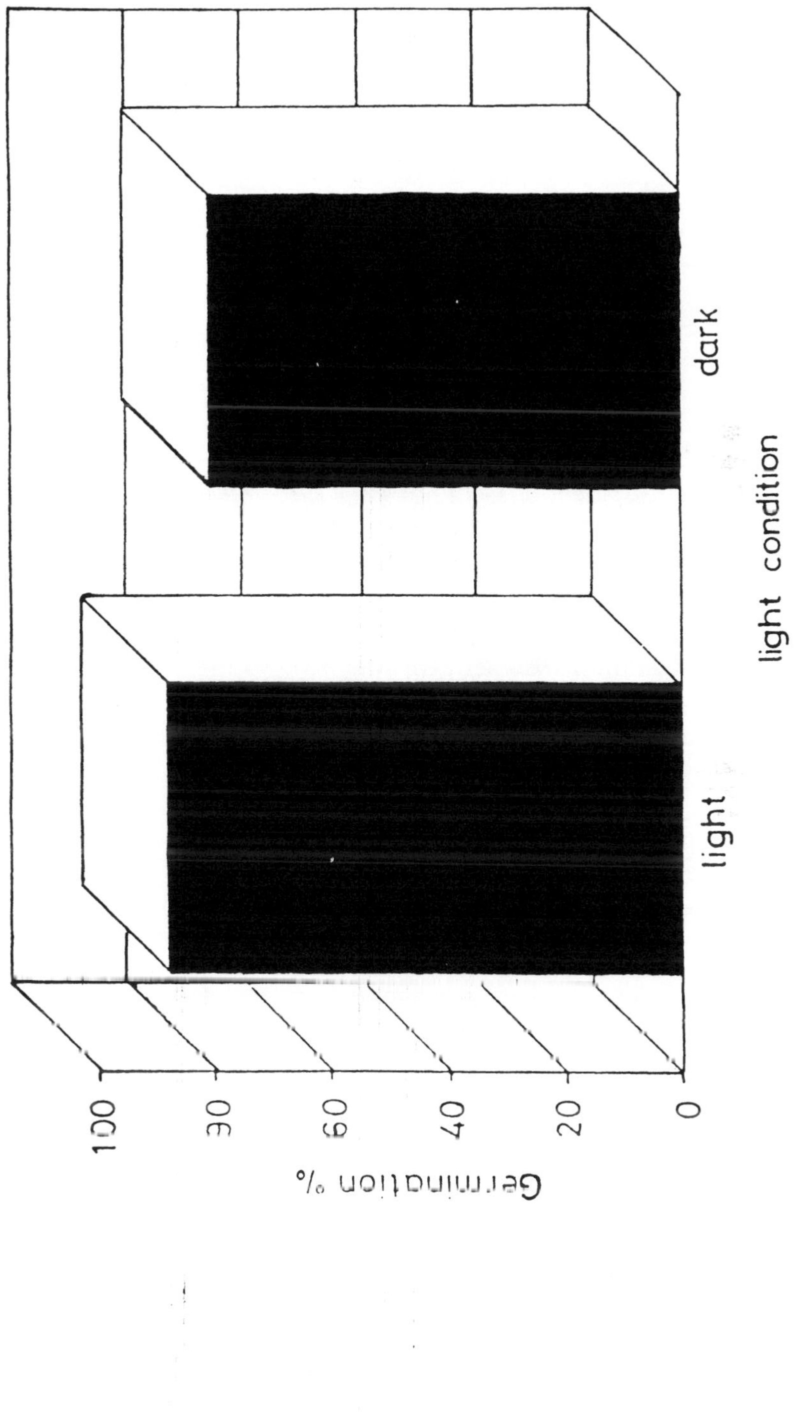

Fig. 4.1: Effect of light on *D. chrysanthum* seed germination

Plate 4: Effect of light and pH on seed germination and seedling growth

1. development of seedlings in *D. chrysanthum* from *in vitro* seed cultures in light and dark conditions.
2. protocorms of *P. spicerianum* from *in vitro* seed cultures kept continuously in dark.
3. protocorms of *P. spicerianum* from *in vitro* seed cultures developing chlorophyll when kept in light 5 days after the commencement of germination in dark.
4. development of seedlings in *D. chrysanthum* from *in vitro* seed cultures at different pH.

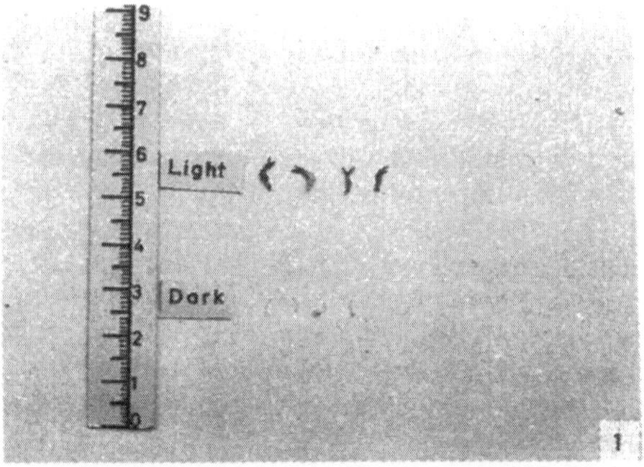
Light
Dark
1

2

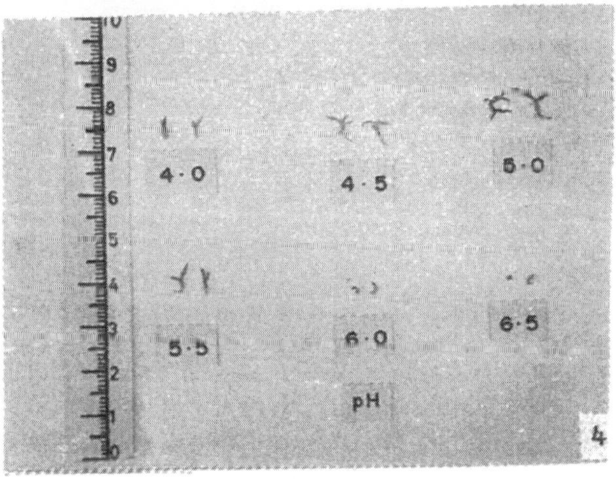
4·0
4·5
5·0
5·5
6·0
6·5
pH
4

RESULTS

Light

D. chrysanthum: Table 4.1 gives data on the effect of light and dark on seed germination and seedling growth. *D. chrysanthum* seeds germinated simultaneously both in light and dark. The germinated seeds under light conditions developed green protocorms but in dark formed only white protocorms. The percentage of seed germination in presence of light was higher than the germination in dark (Table 4.1, Fig. 4.1). A comparison of growth of seedlings obtained from seeds germinated in light and dark conditions, revealed that both leaf and root differentiation occurred in the seedlings obtained from seeds germinated in light but the seedlings kept in dark differentiated only leaf (Table 4.1, Plate 4.1). In such seedlings roots did not differentiate and seedlings died after 90 days, if incubated continuously in dark. However, seedlings differentiated in light conditions survived beyond 90 days.

Table 4.2: Effect of pH on seed germination and seedling growth in *D. chrysanthum* and *P. spicerianum.*

pH	*Commencement of germination (days)*		*Germination per cent*		*Differentiation after 90 days*		*Pigmentation of seedlings*	
	1	*2*	*1*	*2*	*1*	*2*	*1*	*2*
3.0	–	–	0	0	–	–	–	–
4.0	10	35	64.0	32.0	++	+	DG	LG
4.5	10	25	84.5	76.5	+++	+++	DG	DG
5.0	10	25	87.0	79.5	+++	+++	DG	DG
5.5	10	25	77.0	68.0	+++	++	DG	DG
6.0	10	50	55.5	13.0	++	+	MG	LG
6.5	10	–	43.5	0	+	–	MG	–

1 *D. chrysanthum*
2 *P. spicerianum*
\+ protocorm stage
++ seedlings exhibiting leaf differentiation
+++ seedlings exhibiting leaf and root differentiation
LG light green
MG moderately green
DG dark green

P. spicerianum: Effect of light and dark on seed germination and seedling growth is given in Table 4.1. No germination occurred in seeds incubated in light (Fig. 4.2) but the seeds incubated in dark germinated by developing white protocorms on the 25th day. Protocorms kept continuously in dark did not develop chlorophyll *(Plate 4.2),* while those transferred to light developed chlorophyll *(Plate 4.3)* on the 5th day after being transferred to light (Table 4.1). The protocorms kept in dark differentiated only leaf; no root differentiation occurred (Table 4.1). The differentiated seedlings were white and did not survive beyond 90 days (Table 4.1). On the other hand, protocorms transferred to light differentiated both leaf and roots and the seedlings survived even beyond 90 days.

pH

D. chrysanthum: Seed germination and seedling growth were variously influenced by differing pH of the nutrient medium (Table 4.2). No seeds germinated at pH 3.0. Germination of seeds occurred on the 10th day at pH 4.0-6.5. With the increase in pH of the medium germination of seeds increased upto pH 5.0 subsequent to which it declined (Table 4.2, Fig. 4.3). The intensity of decline increased with the increasing pH. Thus at pH 6.5 only 43.5 per cent seeds germinated (Table 4.2). The suitability of pH for seed germination was as follows: 5.0 > 4.5 > 5.5 > 4.0 > 6.0 > 6.5. Only at pH 4.5-5.5 seedlings developed roots (Table 4.2, Plate 4.4). At pH 4.0 and 6.0 seedlings did not develop roots (Table 4.2, Plate 4.4). Seedlings differentiating on medium having pH 4.0-5.5 were dark green in colour but seedlings differentiating on medium having pH 6.0-6.5 were moderately green (Table 4.2).

P. spicerianum : Seed germination and seedling growth were variously influenced by differing pH of the nutrient medium (Table 4.2). No seeds germinated at pH 3.0 and 6.5 (Table 4.2, Fig. 4.4). Germination of seeds occurred earliest at pH 4.5-5.5 (Table 4.2). Seeds germinated on the 35th day at pH 4.0 but germination of seeds occurred on the 50th day at pH 6.0. With the increase in pH of medium seed germination increased upto pH 5.0 subsequent to which it declined (Table 4.2, Fig. 4.4). Thus, at pH 6.0 only 13per cent seeds germinated. The suitability of pH for seed germination was as follows : 5.0 > 4.5 > 5.5 > 4.0 > 6.0. Only at pH 4.5 and 5.0 seedlings developed roots. At pH 5.5 seedlings did not develop roots. However, at pH 4.0 and 6.0 no seedling differentiation occurred (Plate 4.2). Seedlings differentiating on medium having pH 4.5, 5.0 and 5.5 were dark green in colour but seedlings differentiating on medium having pH 4.0 and 6.0 were light green in colour.

DISCUSSION

Arditti (1967) defined germination of seeds in orchids as the swelling of embryos followed by their emergence out of the ruptured seed coat. This is called protocorm stage (Arditti, 1967, 1979; Arditti *et al.*, 1981; Oliva and Arditti, 1984; Mitra, 1986). Further development is considered in terms of appearance of chlorophyll, shoots and roots (Oliva *et al.*, 1984). In the present investigation seeds of *D. chrysanthum* germinated freely to produce chlorophyllous protocorms in the presence of light. But in case of *P. spicerianum* seeds germinated only in dark and produced white protocorms. Thus germinating orchid seeds and developing seedlings vary in their tolerance, requirement and response to light. Similar were the findings of Arditti *et al.*, (1981) and Harvais (1972).

During the present investigation, epiphytic species *D. chrysanthum* germinated in both light and dark. This is in agreement with the earlier findings of Arditti (1967, 1979); Arditti *et al.*, (1981) for epiphytic orchids. However, light proved to be beneficial for germination of seeds into green protocorms, their differentiation into both shoot and root, and survival. Darkness, however, had

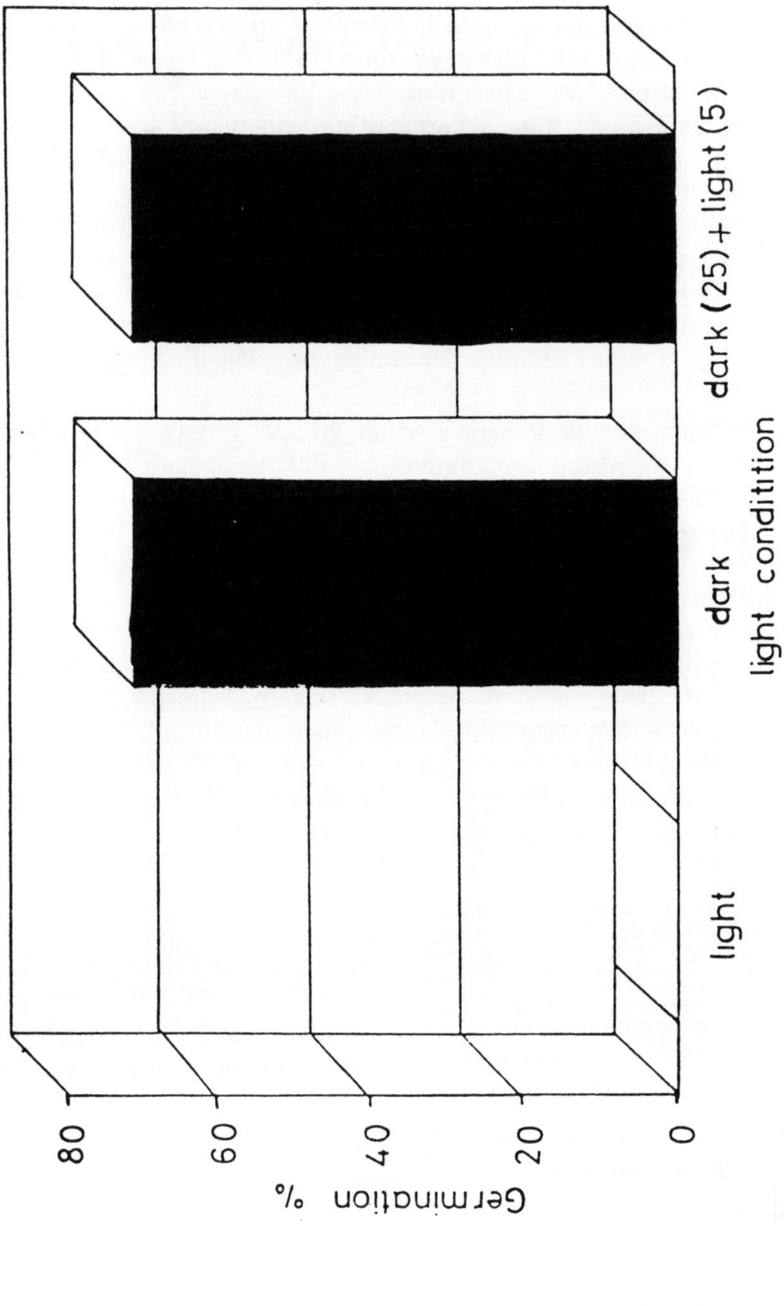

Fig 4.2 : Effect of light on *P. spicerianum* seed germination

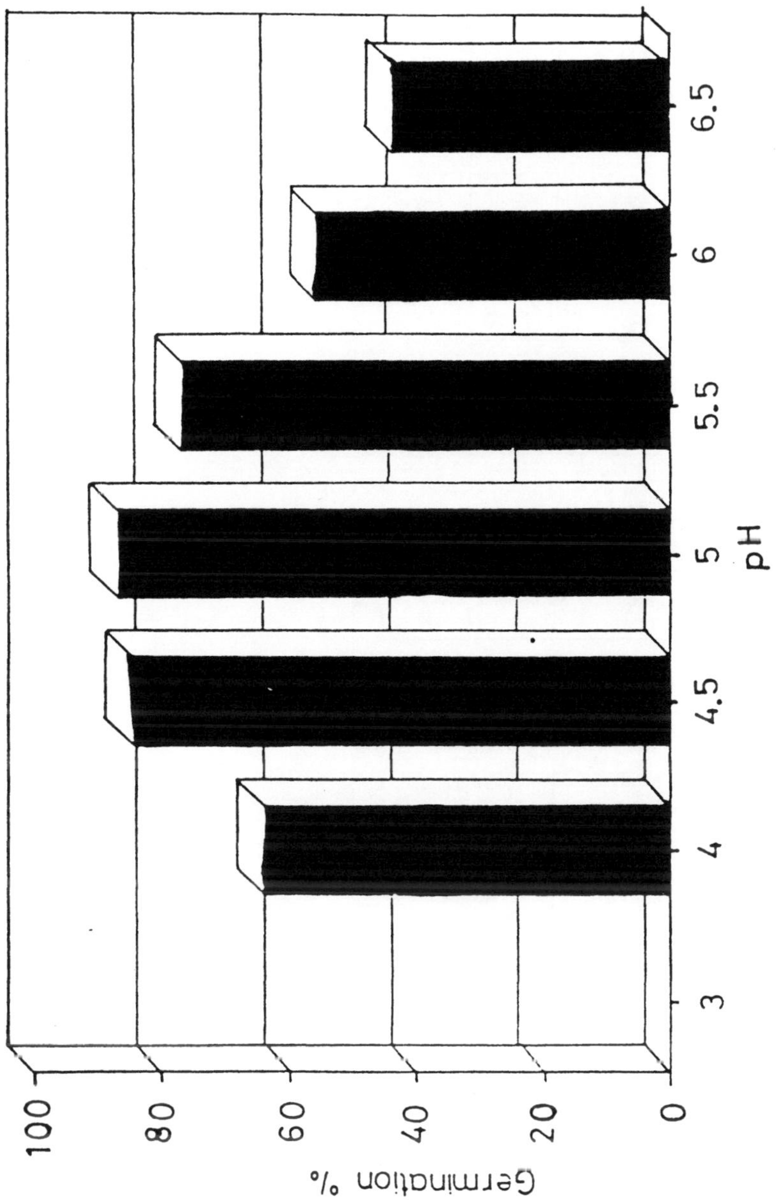

Fig. 4.3 : pH effect on *D. chrysanthum* seed germination

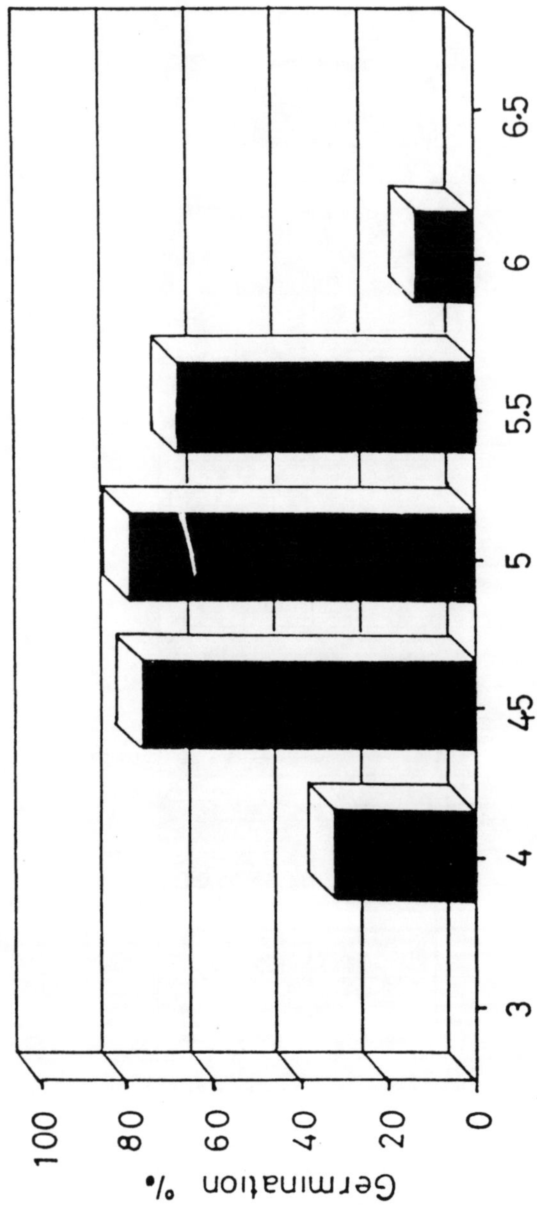

Fig 4.4 : pH effect on *P. spicerianum* seed germination

an inhibitory effect on seed germination and differentiation, pigmentation and survivality of seedlings. For the germination of seeds in the terrestrial species *P. spicerianum* light proved inhibitory. Similar results were also reported in some other terrestrial species such as *Cypripedium reginae* (Harvais, 1972; Stoutamire, 1974) and *Paphiopedilum callosum* (Kano, 1965). Stoutamire (1974) considers that the light-induced inhibition of seed germination in temperate terrestrial orchids is a protective mechanism. The present investigation also supports these conclusions of Stoutamire. In the present investigation light was beneficial for the development of chlorophyll in the protocorms. Unlike the epiphytic species, *D. chrysanthum*, where the germinated seeds were green at the outset, the protocorms of *P. spicerianum*, when transferred to light, turned green with time. This difference in the behaviour of *P. spicerianum* and *D. chrysanthum* is because the seeds of *P. spicerianum* do not germinate in light. The photosynthetic protocorms of *P. spicerianum* differentiate both shoots and roots and reveal a high survival percentage. However, if *P. spicerianum* germinated seeds continued to be incubated in dark only shoots differentiated, and no development of chlorophyll and roots occurred. Ultimately the seedlings of *P. spicerianum* kept in the dark did not survive. Therefore even though the seeds of *P. spicerianum* are to be germinated in dark just after seed germination, the germinated seeds should be transferred to light for proper development and survival of seedlings. Thus, light is, essential for chlorophyll, shoot and root development and for the survival of the seedlings in both the taxa.

pH studies in both the taxa showed no germination or poor germination and growth at pH below 4.0 and above 5.5, suggesting that such low and high pH of the medium are not suitable. Vacin and Went (1949) suggested that inhibition of seed germination below pH 4.0 probably is due to the increased availability of minor elements, such as iron and manganese. Withner (1959) suggests poor growth with higher pH to be due to precipitation of important inorganic nutrients. In the present investigation, moderately acidic medium (pH 5.0) was most suitable for seed germination and seedling growth in both the taxa. Suitability of medium having pH 5.0 could be related to regulating the availability of compounds rather than having a direct effect on growth.

The present observations of an optimal pH are in accord with studies in *Sarcanthus pallidus* (Raghuwanshi *et al.*, 1986) and *D. nobile* (Quednow, 1930 and Ito, 1955). Optimum pH for orchid seed germination has been reported to be 4.8 to 5.2 (Arditti, 1967, 1979; Arditti and Ernst, 1984).

5
Effect of Carbohydrates, Nitrogen Sources, Growth Hormones and Organic Supplements on Seed Germination and Seedling Growth

Introduction

Nutritional studies of plant embryos *in vitro* are of considerable importance in understanding the biochemical ontogeny of embryogenesis in plants (Arditti, 1967; Withner, 1959; Raghavan and Torrey, 1964; Vij *et al.*, 1981). Several workers (Knudson, 1922, 1959; Raghavan and Torrey, 1964; Arditti, 1967, 1979; Ernst, 1967; Pierik and Steegmans, 1972; Mead and Bullard, 1975, 1979; Strauss and Reisinger, 1976; Van Waes and Debergh, 1986; Katiyar *et al.*, 1986; Hew *et al.*, 1987; Sharma and Tandon, 1990; Ernst and Arditti, 1990) have investigated nutritional requirement of orchid embryos.

The earlier studies on the suitability of using sugars as a carbon source for germinating orchid seeds were made before the formation of a symbiotic media (Bernard, 1909). Comparisons between carbon sources became possible only after well defined media were available. As a result, the first comparative study was published only after Knudson's initial publication (Knudson, 1921, 1922). Subsequently many studies were undertaken to identify the carbohydrates capable/ incapable of supporting orchid seed germination (Quednow, 1930; Wynd, 1933; Downie, 1940, 1941, 1943, 1949a, b; Ernst, 1967; Arditti, 1979; Sharma and Tandon, 1990). A wide variety of carbohydrates may act as carbon source for germinating orchid seeds and developing young seedlings. Orchids have been reported to utilize monosaccharide (eg., glucose, fructose); disaccharides (eg., sucrose, cellobiose, maltose, lactose) and polysaccharide (eg., soluble starch) for seed germination (Arditti, 1967, Arditti and Ernst, 1984). Orchids eventhough are able to utilize many different carbohydrates for seed germination they exhibit some preferences. While *Vanilla madagascariensis* require glucose for better germination (Tonnier, 1954); sucrose is good for *Dendrobium* (Israel, 1963) and *Paphiopedilum callosum* (Kano, 1965). The most commonly used carbohydrate for orchid seed and seedling culture is sucrose. But the effect of carbohydrates may vary depending on the concentration (Schaffstein, 1938; Bouriquet, 1947; Kano, 1965; Sharma and Tandon, 1990).

The need for appropriate sources of nitrogen, during germination of orchid seeds, has long been recognised. It has been suggested that an important role of mycorrhizal fungus is to assist with nitrogen nutrition (Magrou, 1944). The inorganic nitrogen sources, ammonium and nitrogen are suitable for seed germination in orchids (Arditti and Ernst, 1948). Both nitrate and ammonia may be utilized by orchids (Curtis and Spoerl, 1948) and are apparently being absorbed at the same rate (Sideris, 1950). Burgeff (1936) reported that ammonium nitrogen is required by *Paphiopedilum* seeds. Ammonium tends to favour the growth of more heterotrophic groups (Arditti and Ernst, 1984). However, a number of species are unable to utilize nitrate during early phase of seed germination (de Bruijne and Debergh, 1974). Development of their ability to utilize nitrate parallels the appearance of nitrate reductase (Arditti and Ernst, 1984). Different orchid genera/species differ in their requirements of ammonia and nitrate (Arditti and Ernst, 1984).

Nearly all amino acids and related substances have been tested in orchid-seed-culture media as supplements or nitrogen sources. Arginine could replace NH_4NO_3 in *Cattleya trianei* cultures (Spoerl, 1948). Phenylalanine, aspartic acid were unable to function as nitrogen sources (Arditti and Ernst, 1984). Glycine inhibited seed germination (Raghavan and Torrey, 1964). Aspartic acid could not serve as a good nitrogen source for some orchid embryos (Arditti and Ernst, 1984).

The discovery of various plant hormones led to their utilization in attempts to promote orchid seed germination and seedling growth. Growth hormones both inhibit and promote seed germination in orchids depending on the type of growth hormone and concentration of growth hormone used (Arditti *et al.*, 1981). Several workers have tried various growth regulators in an attempt to promote seed germination and seedling growth (Arditti, 1967, 1977; Arditti *et al.*, 1982; Arditti and Ernst, 1984; Hadley and Harvais, 1968; Goh, 1970; Hadley, 1970; Fonnesbech, 1972; Pierik and Steegmans, 1972; Harvais, 1973; Strauss and Reisinger, 1976; Flamee, 1978; Nakamura, 1982; Van Waes and Debergh, 1986). Cytokinins usually have no beneficial effect on orchid seed germination but auxins frequently stimulate germination and growth (Arditti, 1967). Traces of auxins have been detected in *Cypripedium* seeds but not in *Dendrobium* and *Calanthe* (Poddubnaya-Arnoldi, 1960; Poddubnaya-Arnoldi and Zinger, 1961). Exogenous application of gibberellins usually has negative effect on the growth of orchid seedlings (Arditti and Ernst, 1984). It seems that either orchid seedlings synthesize the required quantity of gibberellic acid and/or the orchid seeds/seedlings have limited ability to deactivate GA. However, in *Dendrobium nobile* gibberellic acid stimulates seed germination and seedling growth (Miyazaki and Nagamatsu, 1965; Hadley and Harvais, 1968). In *Cymbidium macrorhizon* also gibberellic acid proved beneficial during early stages of germination but suppressed chlorophyll development (Vij and Pathak, 1988).

On the premise that orchid seeds may require supplements of other than a carbon source and some of the known growth factors, various investigators have tested to effects of complex additives on the germination of orchid seeds. Coconut milk stimulated germination of *Paphiopedilum, Vanilla, Cattleya* (Hegarty, 1955). Coconut milk though did not influence germination in *Dendrobium* (Kotomori and Murashige, 1965), 15-25 per cent coconut milk was most effective for protocorms formation in *Cattleya* (Scully, 1967). Incorporation of apple juice in germinating medium induced germination in *Dendrobium* (Kano, 1965). Peptone enhanced germination of *Paphiopedilum insigne, Cymbidium virescence, Cypripedium reginae, Brassolaeliocattleya* and *Dendrobium* (Kano, 1965). Banana homogenate markedly improves germination of seeds in various orchid species (Hey, 1966 a, b). Thus, by adding chemical factors in the nutrient medium, *in vitro* germination of orchid seeds can be manipulated. This chapter therefore deals with the effect of various carbohydrate sources, nitrogen sources, growth hormones and organic supplements on seed germination and subsequent growth of *Dendrobium chrysanthum* and *Paphiopedilum spicerianum*.

MATERIALS AND METHODS

Capsules of *Dendrobium chrysanthum* (330 days old) and *Paphiopedilum spicerianum* (270 days old) were collected. The capsules were surface sterilized by immersion in ethyl alcohol (90%) for a few seconds and flaming. The seeds from surface sterilized capsules were taken out by cutting open the capsules with sterilized scalpel. The seeds of *D. chrysanthum* and *P. spicerianum* were germinated respectively on basal Knudson C (Kn C) and Thomale GD (Th GD) medium with various chemical treatments. Preparation of the two medium is described in chapter 3. Approximately 50 seeds were sown per culture tube containing 20 ml medium. Per treatment ten tubes were used. The cultures of *D. chrysanthum* were kept under continuous light (2,500-3,000 lux). But cultures of *P. spicerianum* were first kept in dark and five days after the commencement of germination were transferred to light. The cultures were maintained at 25 ± 1°C temperature.

For the study of the effects of various chemical treatments on *in vitro* seed germination and its subsequent growth, the modifications made in the respective medium were:

(a) *Carbohydrate sources:* Basal medium devoid of any carbohydrate sources served as control. To find out the effect of various carbohydrates (glucose, fructose, sucrose, lactose and soluble starch) on seed germination, these were incorporated in the basal medium one at a time. The concentrations used were 1-5 per cent.

(b) *Nitrogen sources:* Basal medium devoid of any nitrogen source served as control. To investigate the effect of various inorganic nitrogen (ammonium nitrate, calcium nitrate, sodium nitrate, ammonium sulphate and potassium nitrate) and organic nitrogen sources (amino acids, L-arginine, L-aspartic acid, L-phenylalanine, glycine) on the germination of the seeds, 0.1 g l^{-1} – 2.0 g l^{-1} of these were incorporated in the basal medium individually to determine the most suitable nitrogen source.

(c) *Growth hormones:* Basal medium was taken as control medium. To determine the effect of various growth hormones on seed germination and subsequent seedling growth, 0.1 – 10.0 mg l^{-1} concentrations of various cytokinins (Kinetin, KN, and 6-benzylaminopurine, BAP), auxins (α-naphthylacetic acid, NAA; indole acetic acid, IAA; and 2,4-dichlorophenoxy acetic acid, 2, 4-D) and gibberellic acid (GA_3) were incorporated, one at a time, in the control medium.

(d) *Organic supplements:* Basal medium was treated as the control medium. Various organic supplements such as banana homogenate (BM), 50 gl^{-1};

apple juice, 50 ml l^{-1}; coconut milk (CM), 150 ml l^{-1}; peptone, 100 mg l^{-1} and activated charcoal (AC), 100 mg l^{-1} were added to the germinating medium to determine their effect on *in vitro* seed gemination and seedling growth.

Preparation of banana homogenate: 50 g of ripe banana fruit was homogenised in a blender and added to the medium. The final volume was made to 1 lit. with distilled water.

Preparation of apple juice: 100 g of ripe apple fruit cut into pieces was placed in a blender and homogenised until apple pieces became pulpy. The apple pulp was strained through a cheese cloth to obtain the juice. 50 ml of the apple juice was used per lit. of medium.

Coconut milk: This is the liquid endosperm of *Cocos nucifera*. Green coconut were taken, cut open to get the liquid. 150 ml of the liquid endosperm was used per lit. of medium.

Data Recording

In both the genus, *Dendrobium* and *Paphiopedilum* time taken to commence germination was recorded. Percentage seed germination was recorded five days after the commencement of germination. On completion of the experiment, on the 90th day, differentiation of seedlings (protocorm stage; differentiation of only leaf; differentiation of both leaf and roots) was recorded. Pigmentation of the seedlings was also recorded on the 90th day.

RESULTS

Dendrobium chrysanthum

Carbohydrates: Effect of various concentrations (1-5%) of different carbohydrate sources on *Dendrobium chrysanthum* seed germination and seedling growth are given in Table 5.1.

Control: *Dendrobium* seeds germinated on the 10th day on medium devoid of carbohydrate source. The germination percentage recorded was 26%. But the germinated seeds did not differentiate beyond the protocorm stage. The protocorms were pale green in colour.

Treated: Seed germination occurred on the 10th day on medium containing 1-5 per cent of glucose, fructose, sucrose, lactose and 1-2 per cent of soluble starch. But seeds did not germinate upto the 90th day on medium having 3-5 per cent soluble starch.

Glucose, fructose, sucrose and lactose enhanced germination over control. Soluble starch, however, inhibited germination. Amongst glucose, fructose, lactose and soluble starch maximum germination was recorded with their minimum incorporation (1%) (Fig. 5.1, 5.2, 5.4 & 5.5), the decrease in germination was proportional to the increase in the concentration of the respective carbohydrates. Incorporation of sucrose into the medium stimulated germination of seeds (Table 5.1, Fig. 5.3). The germination increased with the increase in the concentration of sucrose upto 2 per cent and declined thereafter. Amongst the various carbohydrate sources 2 per cent sucrose induced maximum germination of seeds. The effectiveness of various carbohydrate sources in causing germination of seeds was in the following order: sucrose, glucose, fructose and lactose.

Table 5.1: Effect of different concentrations of various carbohydrate source on seed germination, seedling differentiation and pigmentation of the seedlings in *D. chrysanthum*.

Character	*Control*	*Carbohydrates (%)*																								
		Glucose					*Fructose*					*Sucrose*					*Lactose*					*Soluble Starch*				
		1	*2*	*3*	*4*	*5*	*1*	*2*	*3*	*4*	*5*	*1*	*2*	*3*	*4*	*5*	*1*	*2*	*3*	*4*	*5*	*1*	*2*	*3*	*4*	*5*
Time taken to commence germination (days)	10	10	10	10	10	10	10	10	10	10	10	10	10	10	10	10	10	10	10	10	10	10	10	–	–	–
Germination per cent	26	75	69	64	59	49	73	67	63	61	56	70	87	70	56	52	62	55	41	31	30	23	12	0	0	0
Differentiation after 90 days	+	+++	++	++	++	++	+++	+++	+++	++	++	+++	+++	+++	++	++	++	++	++	++	++	++	++	–	–	–
Pigmentation of seedlings	PG	MG	MG	MG	MG	MG	MG	MG	MG	MG	MG	MG	MG	MG	MG	MG	LG	LG	LG	LG	LG	LG	LG	–	–	–

+ protocorm stage
++ seedlings exhibiting leaf differentiation
+++ seedlings exhibiting leaf and root differentiation
PG pale green
MG moderately green
LG light green

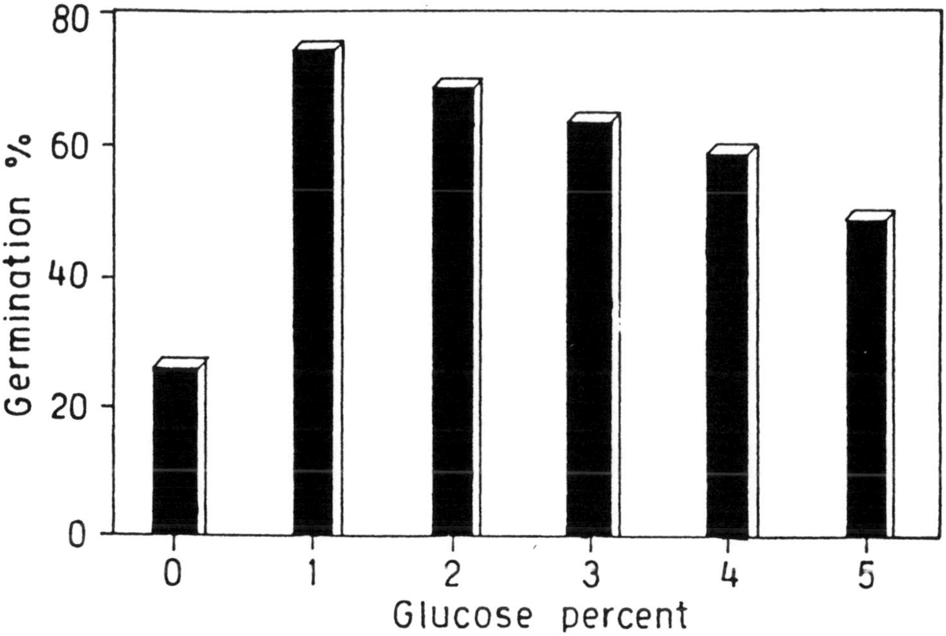

Fig. 5.1 : Glucose effect on *D. chrysanthum* seed germination

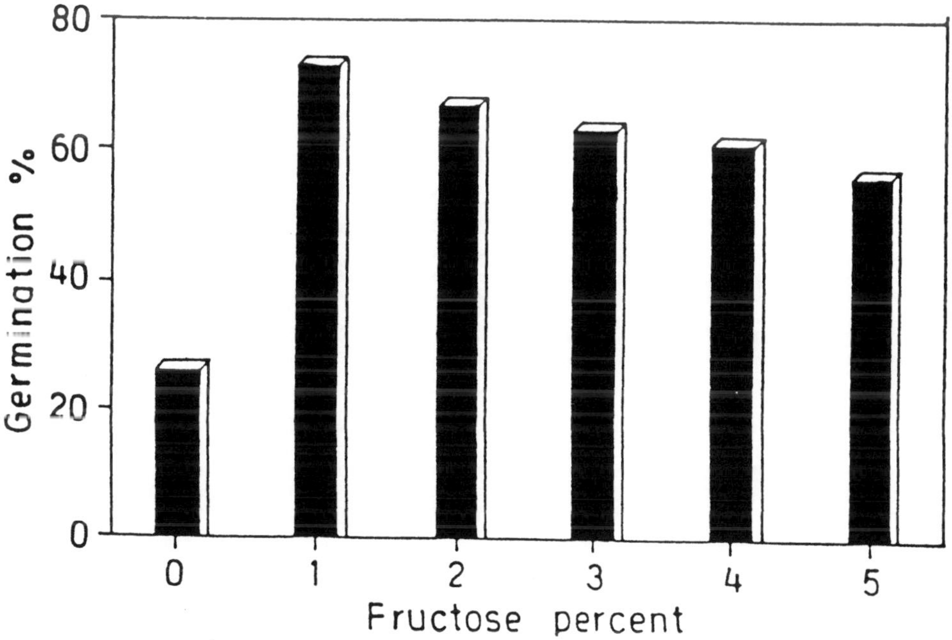

Fig. 5.2 : Fructose effect on *D. chrysanthum* seed germination

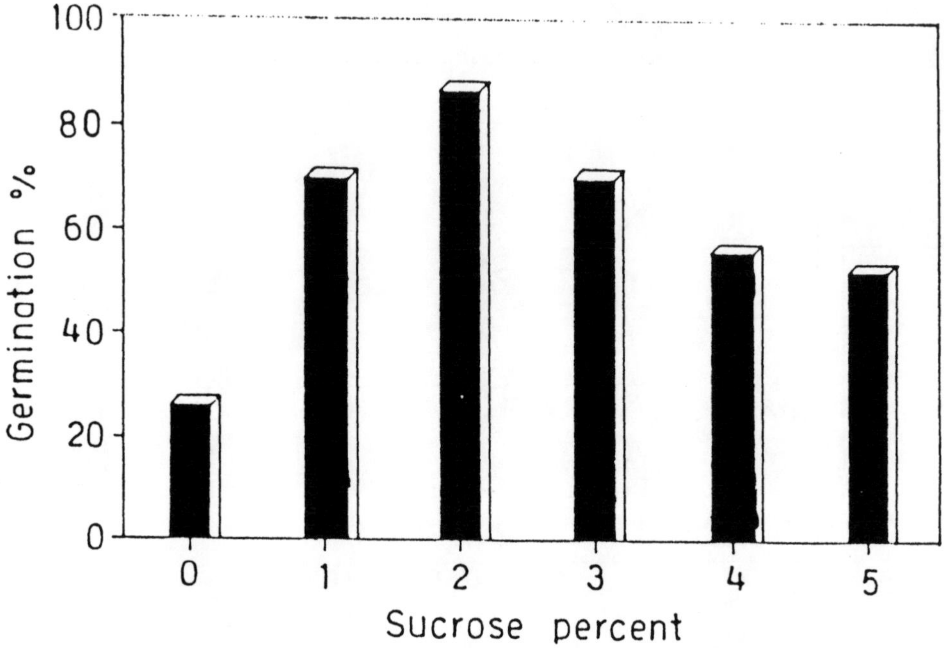

Fig 5.3 : Sucrose effect on *D. chrysanthum* seed germination

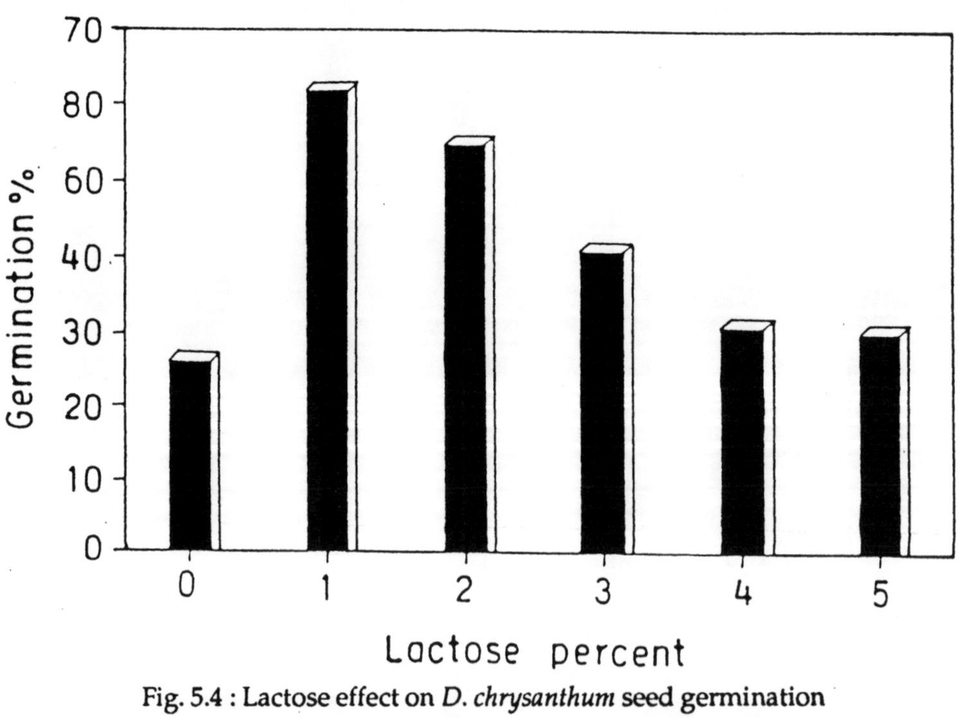

Fig. 5.4 : Lactose effect on *D. chrysanthum* seed germination

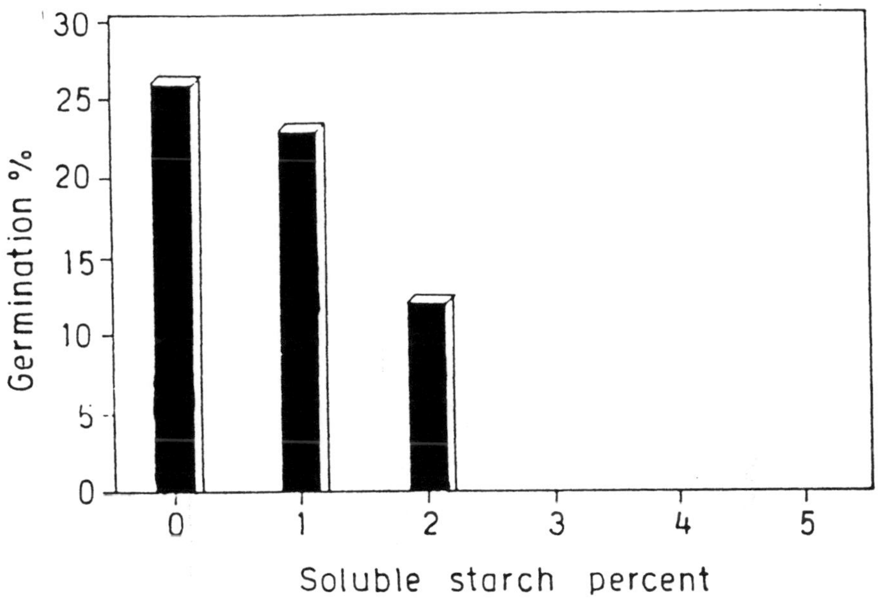

Fig 5.5 : Soluble starch effect on *D. chrysanthum* seed germination

The seedlings developing on medium having various carbohydrate sources differentiated leaf except on medium having 3-5 per cent soluble starch (Table 5.1, *Plate 5.1–5.5*). Differentiation of roots in the seedlings was recorded on medium having either 1 per cent glucose or 1-3% fructose/sucrose (Table 5.1, *Plate 5.1-5.3*).

The colour of seedlings grown on medium incorporated with glucose/fructose/sucrose was moderately green whereas seedlings of light green colour developed on medium supplemented with either lactose or soluble starch.

Nitrogen sources

Amongst nitrogenous sources (L-arginine, L-aspartic acid, L-phenylalanine, glycine, ammonium nitrate, sodium nitrate, calcium nitrate, potassium nitrate and ammonium sulplate), germination of seeds and seedling differentiation occurred only on medium supplemented with inorganic nitrogen sources (ammonium nitrate, sodium nitrate, calcium nitrate, potassium nitrate and ammonium sulphate) (Table 5.2).

Control: *Dendrobium* seeds germinated on the 10th day on medium devoid of any nitrogen source and altogether 26% seeds germinated. The germinated seeds did not differentiate beyond the protocorm stage which were light green in colour (Table 5.2).

Treated: Germination of seeds occurred on 0.1-2.0 gl^{-1} NH_4 NO_3 and Ca $(NO_3)_2$ supplemented medium and 0.1- 1.0 g 1^{-1}, $NaNO_3$, KNO_3 and NH_4 SO_4 supplemented medium, respectively (Table 6.2). No germination of seeds was evident in the presence of 2 g 1^{-1} $NaNO_3$/KNO_3/NH_4SO_4 (Table 5.2).

Plate 5: Effect of different concentrations of various carbohydrates (Fig. 1-5) and growth regulators (Fig. 6-11) on seedling differentiation in *D. chrysanthum*

1. glucose
2. fructose
3. sucrose
4. lactose
5. soluble starch
6. kinetin
7. BAP
8. NAA
9. IAA
10. 2, 4–D
11. GA_3

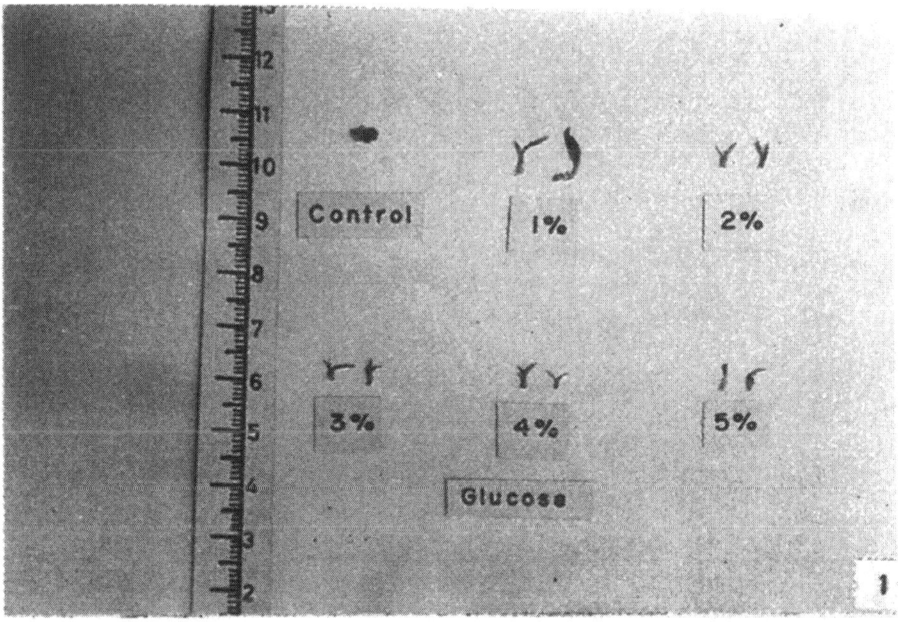
Control
1%
2%
3%
4%
5%
Glucose
1

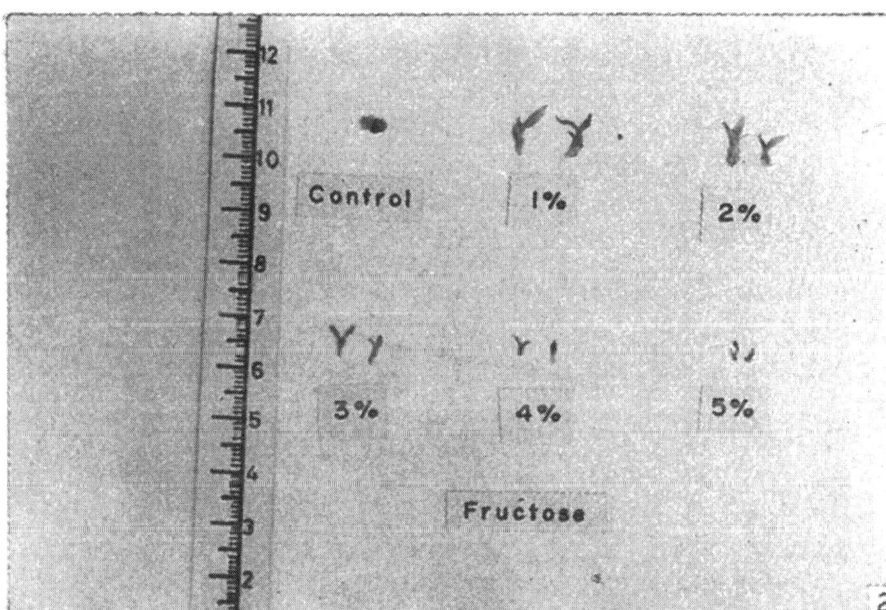
Control
1%
2%
3%
4%
5%
Fructose
2

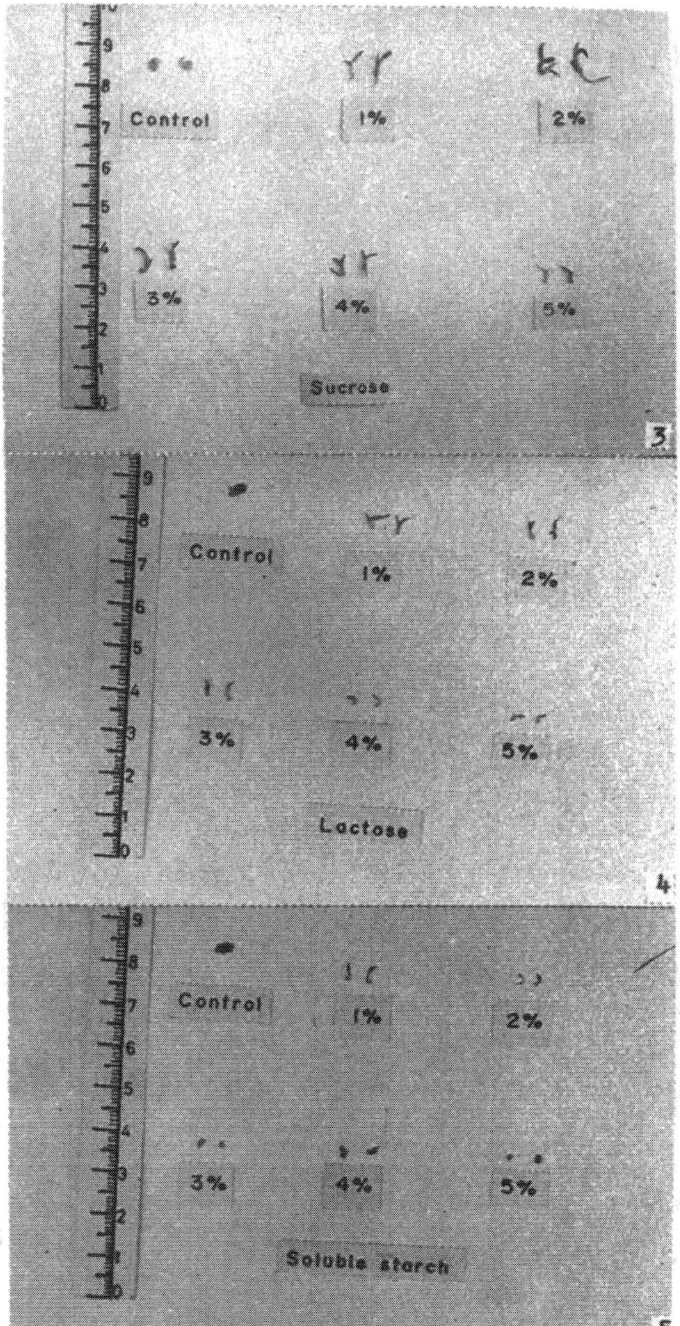
Control
1%
2%
3%
4%
5%
Sucrose
3
Control
1%
2%
3%
4%
5%
Lactose
4
Control
1%
2%
3%
4%
5%
Soluble starch
5

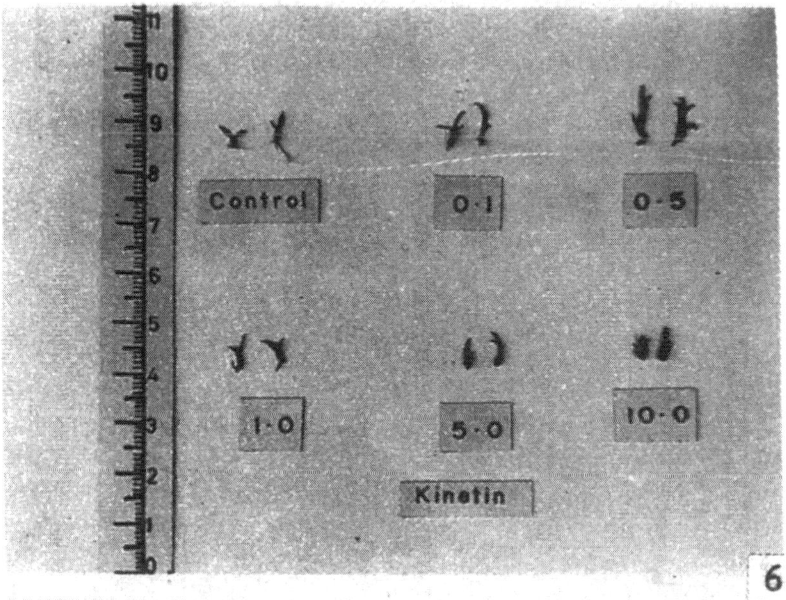

6

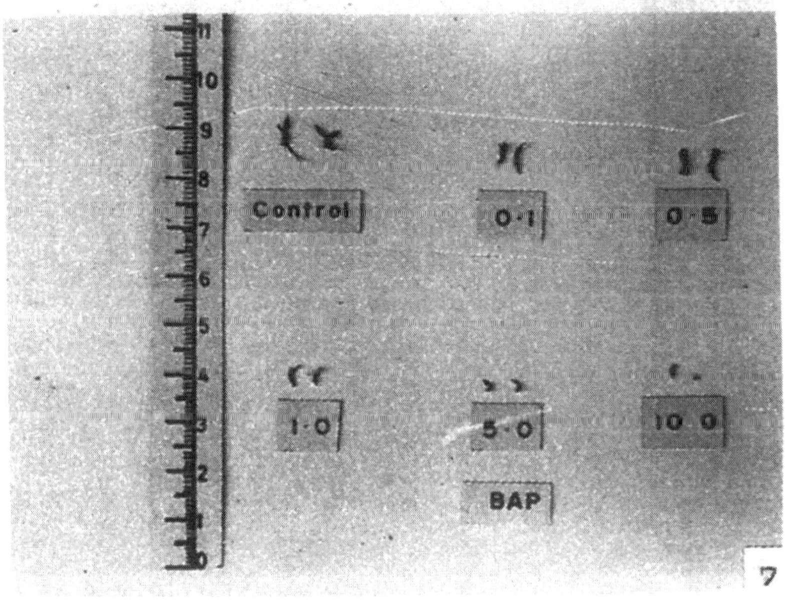

7

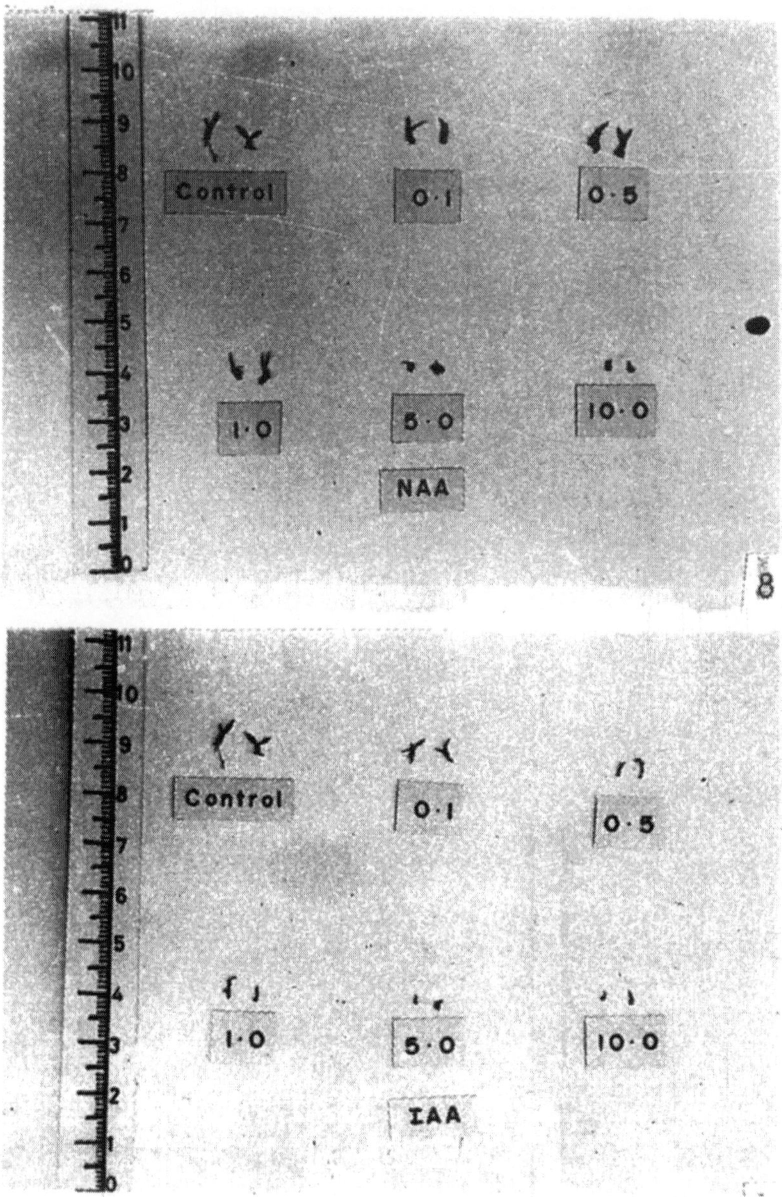
Control
0·1
0·5
1·0
5·0
10·0
NAA
8
Control
0·1
0·5
1·0
5·0
10·0
IAA

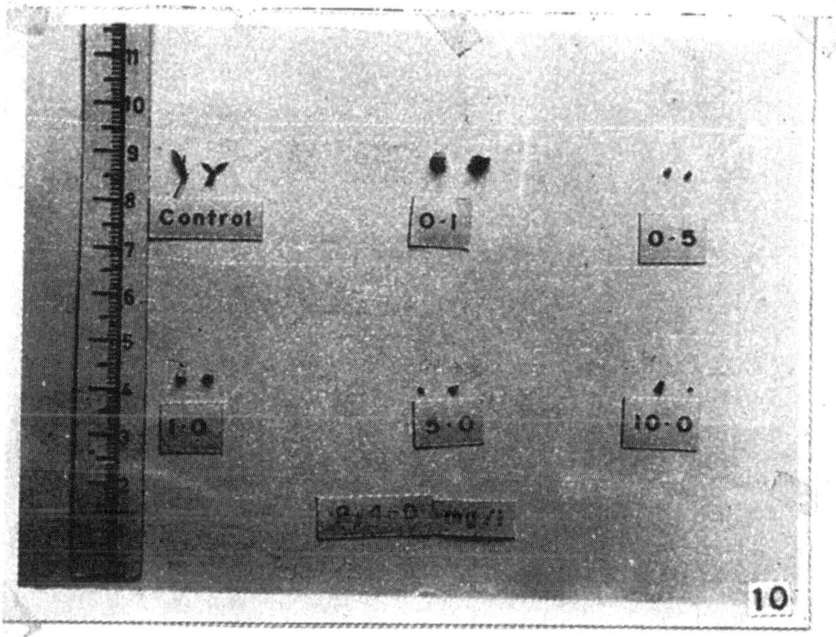
Control
0·1
0·5
1·0
5·0
10·0
10

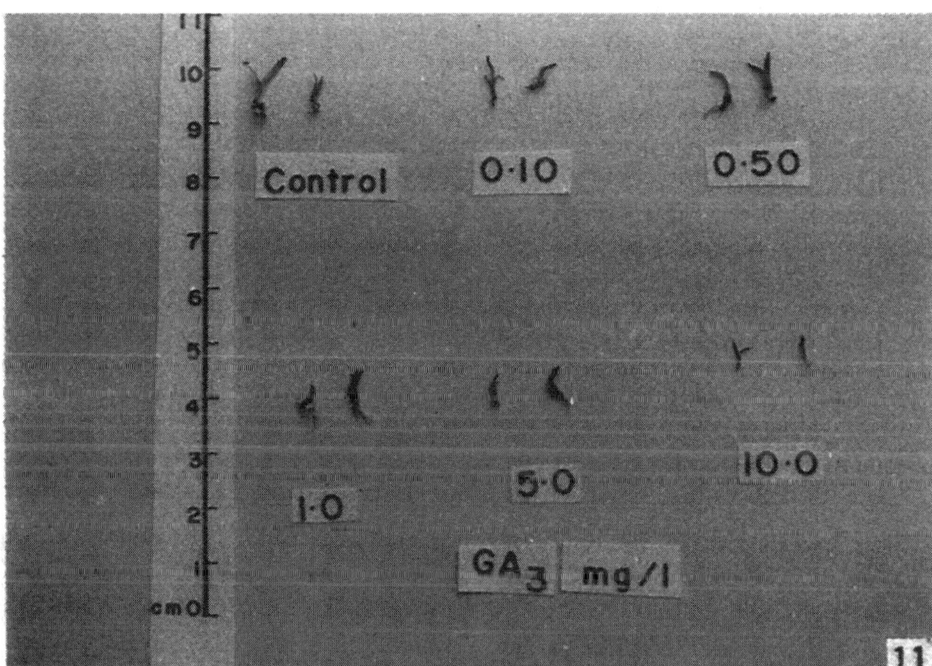
Control
0·10
0·50
1·0
5·0
10·0
GA3 mg/l
11

Table 5.2 : Effect of different concentrations of various inorganic nitrogen source on seed germination and seedling growth in *D. chrysanthum*

Character	*Control*	*Inorganic nitrogens ($g\ l^{-1}$)*																			
		Ammonium nitrate				*Sodium nitrate*				*Calcium nitrate*				*Potassium nitrate*				*Ammonium sulphate*			
		0.1	*0.5*	*1.0*	*2.0*	*0.1*	*0.5*	*1.0*	*2.0*	*0.1*	*0.5*	*1.0*	*2.0*	*0.1*	*0.5*	*1.0*	*2.0*	*0.1*	*0.5*	*1.0*	*2.0*
Time taken to commence germination (days)	10	10	10	10	10	10	10	20	–	10	10	10	10	10	10	10	–	10	10	10	–
Germination per cent	26	28	33.5	24	16	25	28	12	0	49	63	72	50	32	41	36	0	19	26	16	0
Differentiation after 90 days	+	+++	+++	+++	+	++	++	+	–	+++	+++	+++	++	+++	+++	+++	–	++	++	++	–
Pigmentation of seedlings	LG	MG	MG	MG	MG	LG	LG	LG	–	MG	MG	MG	MG	MG	MG	MG	–	MG	MG	MG	–

\+ protocorm stage

++ seedlings exhibiting leaf differentiation

+++ seedlings exhibiting leaf and root differentiation

MG moderately green

LG light green

Commencement of seed germination occurred on the 10th day on medium supplemented with various inorganic nitrogen sources except on medium having $NaNO_3$ 1.0 g 1^{-1}. In this treatment seed germination commenced on the 20th day (Table 5.2).

Seed germination increased with the increasing concentration of NH_4 NO_3. Maximum germination (33.5%) was recorded on medium incorporated with 0.5 g 1^{-1}, NH_4 NO_3. Higher concentration of NH_4NO_3 (1.0-2.0 g 1^{-1}) decreased germination (Table 5.2, Fig. 5.6). A similar trend was recorded wth $NaNO_3$, KNO_3 and NH_4 SO_4. With incorporation of Ca $(NO_3)_2$ in the medium germination percentage increased gradually upto 1.0 g 1^{-1} and thereafter decreased gradually with the increasing concentrations of Ca $(NO_3)_2$ (Table 5.2, Fig. 5.6). Interestingly Ca $(NO_3)_2$ proved to be the best N_2 source in comparison to other inorganic nitrogen sources. The suitability of inorganic nitrogen sources for inducing seed germination was in the following order: Ca $(NO_3)_2$, NH_4 NO_3, $NaNO_3$ and NH_4 SO_4 (Table 5.2, Fig. 5.6).

The seedlings differentiating on medium having lower concentration 0.1–1.0 g 1^{-1} of NH_4 NO_3, Ca $(NO_3)_2$ and KNO_3 differentiated both leaf and root (Table 5.2). At higher concentration of NH_4 NO_3/$NaNO_3$ (2.0 g 1^{-1}) although seedlings did not differentiate beyond the protocorm stage, the seedlings developing on medium supplemented with 2.0 g 1^{-1} Ca $(NO_3)_2$ differentiated leaf (Table 5.2). Seedlings formed on medium having NH_4 SO_4 (0.1–1.0 g 1^{-1}) exhibited only leaf differentiation. Seedlings growing on all the inorganic nitrogen sources were moderately green in colour except for seedlings growing on $NaNO_3$ supplemented medium which were light green in colour (Table 5.2).

Growth Hormones

The effect of growth hormones on seed germination and subsequent growth is given in Table 5.3.

Control: On medium devoid of growth hormones, *Dendrobium* seeds germinated on the 10th day, the germination percentage recorded was 88 (Table 5.3, Fig. 5.7, 5.8). 90 days old seedlings differentiated both leaves and roots. The differentiated seedlings were green in colour.

Treated: Various concentrations of KN; BAP; NAA; IAA; 2, 4–D and GA_3 used did not alter the time required for commencement of seed germination. The only exception being 2, 4–D (10.0 mg 1–1) which did not cause seed germination upto the 90th day. All growth hormones in the present study inhibited seed germination and the inhibition increased with the increasing concentration of growth hormone (Table 5.3, Fig. 5.7, 5.8).

Both leaf and root differentiation remained unaffected in seedlings growing on medium having 0.1–1.0 mg 1^{-1} KN/BAP, 0.1 mg 1^{-1} NAA/IAA and 0.1–5.0 mg 1^{-1} GA_3 (Plate 5.6–5.9 & 5.11). At higher concentration of the respective growth hormones seedlings exhibited only leaf differentiation. Seedlings, growing on 2, 4–D supplemented medium remained in the protocorm stage (Table 5.3, Plate 5.10).

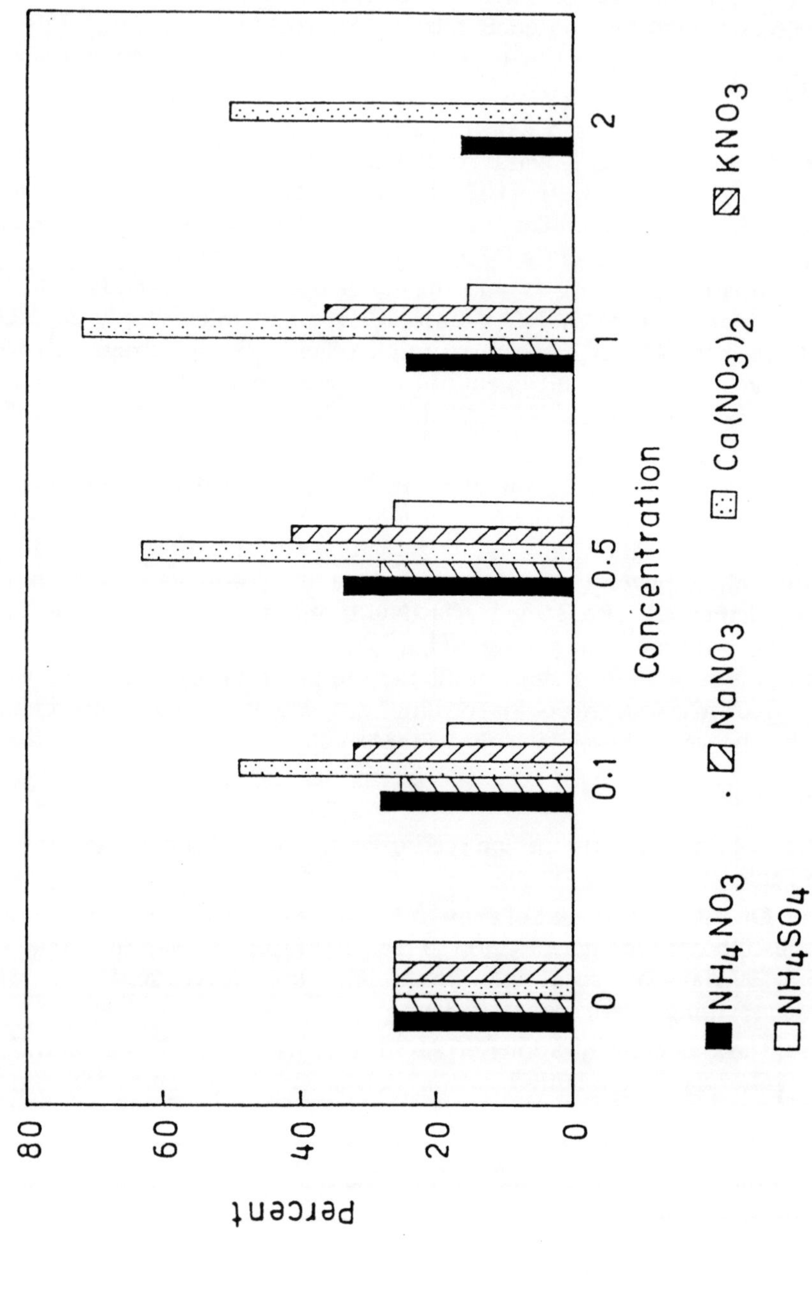

Fig 5.6 : Inorganic nitrogen effect on *D. chrysanthum* seed germination

Table 5.3 : Effect of different concentrations of various growth hormones on seed germination, seedling differentiation and pigmentation of the seedlings in *D. chrysanthum*.

Character	*Control*	*Growth hormones (mg 1⁻¹)*																													
		KN					*BAP*					*NAA*					*IAA*					*2, 4–D*					*GA_3*				
		0.1	*0.5*	*1.0*	*5.0*	*10.0*	*0.1*	*0.5*	*1.0*	*5.0*	*10.0*	*0.1*	*0.5*	*1.0*	*5.0*	*10.0*	*0.1*	*0.5*	*1.0*	*5.0*	*10.0*	*0.1*	*0.5*	*1.0*	*5.0*	*10.0*	*0.1*	*0.5*	*1.0*	*5.0*	*10.0*
Time taken to commence germination (days)	10	10	10	10	10	10	10	10	10	10	10	10	10	10	10	10	10	10	10	10	10	10	10	10	10	-	10	10	10	10	10
Germination percent	88	81	83	68	59	59	64	62	61	48	48	66	58	50	44	36	65	56	52	49	28	34	24	21	15	0	62	60	56	54	46
Differentiation after 90 days	+++	+++	+++	+++	++	++	-++	+++	+++	++	++	+++	++	++	++	++	+++	++	++	++	++	+	+	+	+	-	+++	+++	+++	+++	++
Pigmentation of seedlings	MG	DG	DG	DG	DG	DG	DG	DG	DG	MG	MG	MG	MG	MG	MG	LG	MG	MG	MG	MG	LG	W	W	W	W	-	DG	DG	DG	DG	MG

+ protocorm stage
++ seedlings exhibiting leaf differentiation
+++ seedlings exhibiting leaf and root differentiation
DG dark green
LG light green
W white

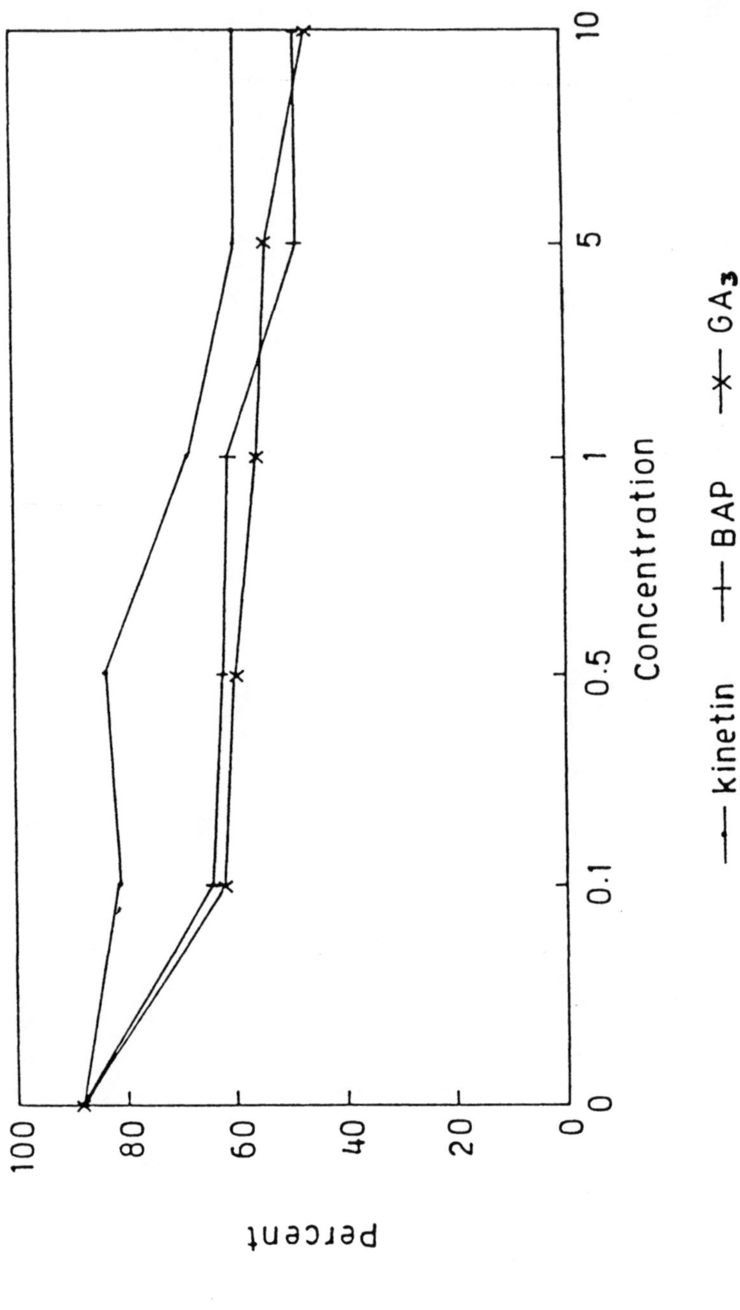

Fig. 5.7 : Cytokinin & GA_3 effect on *D. chrysanthum* seed germination

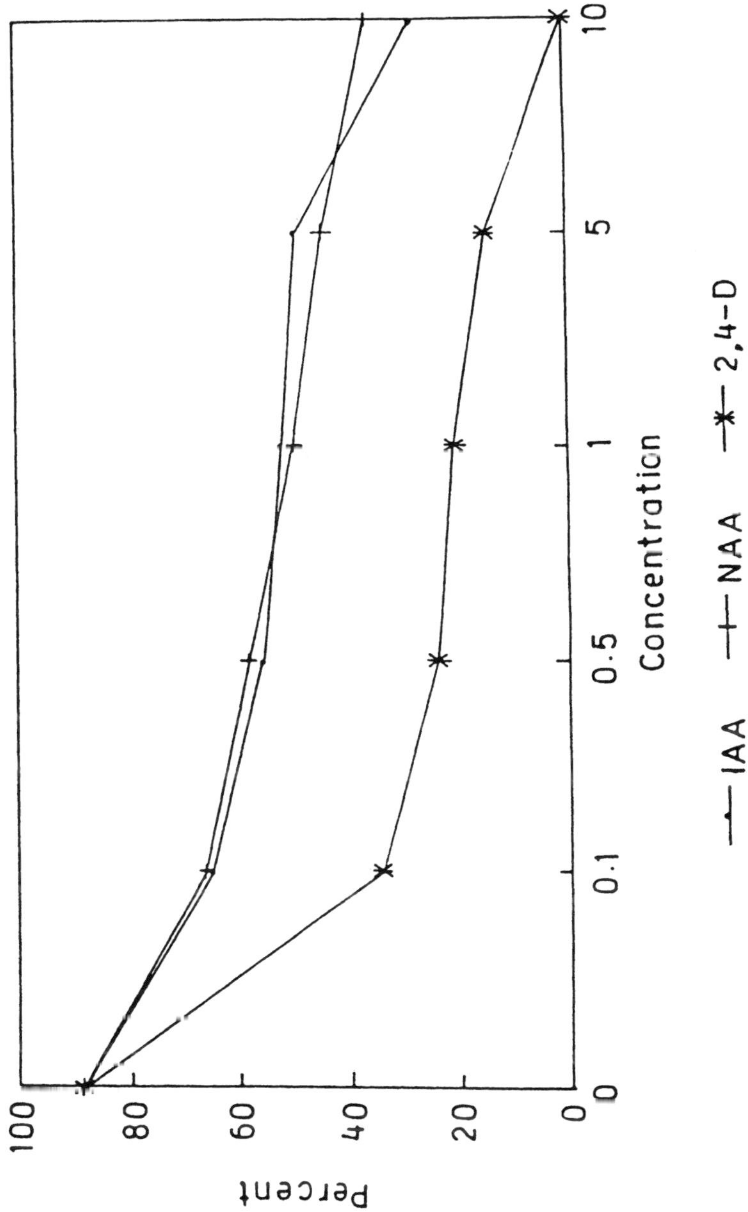

Fig 5.8 : Auxin effect on *D. chrysanthum* seed germination

The seedlings developing on medium incorporated with 0.1–10.0 mg 1^{-1} KN, 0.1–1.0 mg 1^{-1} BAP, 0.1–5.0 mg 1^{-1} GA_3, compared to control were darker green in colour. Seedlings developing on 5.0–10.0 mg 1^{-1} BAP, 0.1-0.5 mg 1^{-1} NAA/IAA and 10.0 mg 1^{-1} GA_3 containing medium were moderately green in colour. The seedlings developing on 10.0 mg 1^{-1} NAA/IAA incorporated medium were light green in colour. On medium having 2, 4–D the protocorms remained white (Table 5.3).

Organic supplements: The effect of organic supplements on seed germination and seedling growth is given in Table 5.4.

Control: On medium devoid of any organic supplements, *Dendrobium* seeds germinated on the 10th day. The degree of germination was 88.5 per cent. 90 days old seedlings differentiated both leaves and roots and were green in colour (Table 5.4).

Treated: Except for CM all other organic supplements (banana homogenate, apple juice, peptone and activated charcoal) delayed the commencement of germination compared to control (Table 5.4).

Table 5.4 : Effect of various organic supplements on seed germination and seedling growth in *D. chrysanthum.*

Character	*Control*	*Organic supplements*				
		Coconut Milk	*Banana Homogenate*	*Apple Juice*	*Peptone*	*Activated Charcoal*
Time taken to commence germination (days)	10	10	20	20	20	20
Germination per cent	88.5	65	55	32	30	65
Differentiation after 90 days	+++	+++	+++	++	++	+++
Pigmentation of seedlings	MG	MG	PG	PG	LG	DG

++ seedlings exhibiting leaf differentiation

+++ seedlings exhibiting leaf and root differentiation

PG pale green DG dark green

LG light green

MG moderately green

Organic supplements inhibited seed germination and the decrease was in the following order: CM = activated charcoal > banana homogenate > apple juice > peptone.

The seedlings formed on medium containing CM, banana homogenate and activated charcoal exhibited differentiation of both leaf and roots. But seedlings developing on medium having either apple juice or peptone differentiated only leaf (Table 5.4).

Addition of activated charcoal to the medium developed dark green seedlings. The seedlings growing on CM incorporated medium were moderately green while pale green seedlings developed on medium having banana homogenate/ apple juice. Light green seedlings developed on peptone supplemented medium (Table 5.4).

Table 5.5 : Effect of different concentrations of various carbohydrate source on seed germination, seedling differentiation and pigmentation of the seedlings in *P. spicerianum.*

Character	*Control*	*Carbohydrates (%)*																								
		Glucose					*Fructose*					*Sucrose*					*Lactose*					*Soluble Starch*				
		1	*2*	*3*	*4*	*5*	*1*	*2*	*3*	*4*	*5*	*1*	*2*	*3*	*4*	*5*	*1*	*2*	*3*	*4*	*5*	*1*	*2*	*3*	*4*	*5*
Time taken to commence germination	–	25	25	25	25	–	30	30	30	50	–	25	25	30	50	–	–	–	–	–	–	-	-	-	-	-
Germination percent	0	25.5	72.5	75	45	0	28	43.5	62	39.5	0	60	78.5	71.5	38	0	0	0	0	0	0	0	0	0	0	0
Differentiation after 90 days	–	++	++	+–	+	–	+	++	++	+	–	++	+++	++	++	–	–	–	–	–	–	–	–	–	–	–
Pigmentation of seedlings	–	PG	LG	MG	W	–	PG	LG	MG	W	–	MG	DG	DG	MG	–	–	–	–	–	–	–	–	–	–	–

+ protocorm stage
++ seedlings exhibiting leaf differentiation
+++ seedlings exhibiting leaf and root differentiation
PG pale green
MG moderately green
LG light green
DG dark green
W white

Paphiopedilum spicerianum

Carbohydrates: Effects of various concentrations (1-5%) of different carbohydrate source on *Paphiopedilum spicerianum* seed germination and seedling growth are given in Table 5.5.

Control: *Paphiopedilum* seeds do not germinate on medium devoid of any carbohydrate source (Table 5.5, Fig. 5.9).

Treated: Seeds germinate earliest (25th day) on the medium containing 1-4% of glucose and 1–2 per cent of sucrose. Whereas seeds sown on medium containing 1-3% of fructose and 3 per cent sucrose germinated on the 30th day. The germination of seeds was evident on the 50th day on medium having 4 per cent fructose/sucrose. No seeds germinated on medium containing 5 per cent glucose/fructose/sucrose and 1–5 per cent lactose/soluble starch (Table 5.5).

The degree of seed germination on glucose, fructose and sucrose supplemented medium increased with the increasing concentration upto a level. Maximum seed germination occurred on medium supplemented with 3 per cent glucose/fructose and 2 per cent sucrose (Table 5.5, Fig. 5.9). Further increase in the concentration of respective carbohydrate diminishes seed germination. Amongst the various carbohydrate sources 2 per cent sucrose induced maximum germination. The relative effectiveness of various carbohydrate sources in causing germination of seeds was in the following order: sucrose, glucose and fructose (Table 5.5).

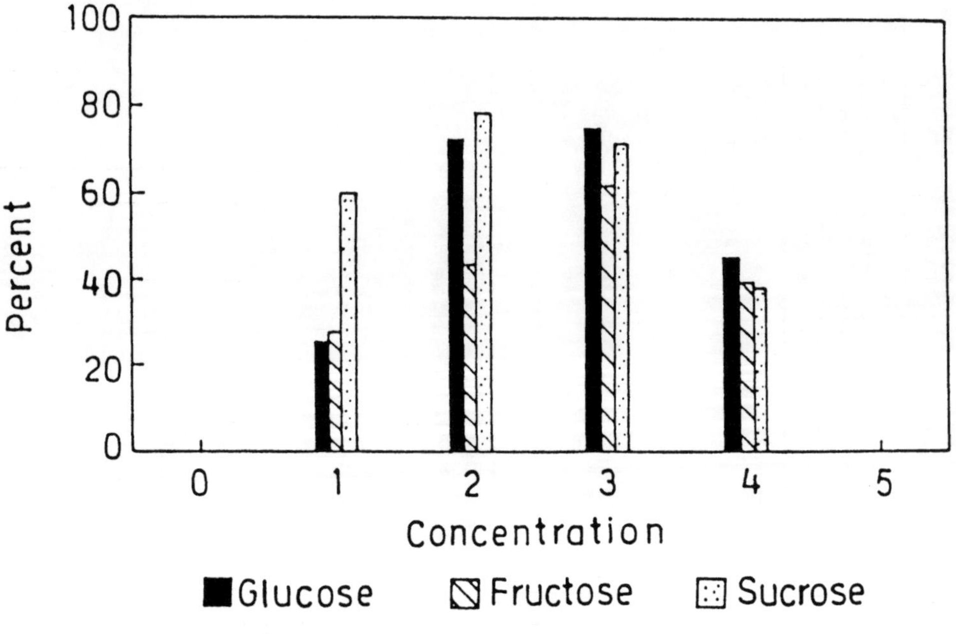

Fig. 5.9 : Carbohydrate effect on *P. spicerianum* seed germination

The seedlings developing on medium having 1-3 per cent glucose, 2–3 per cent fructose, and 1 per cent and 3–4 per cent sucrose differentiated leaf but seedlings grown on medium supplemented with 4 per cent glucose and 1 and 4% fructose did not differentiate beyond protocorm stage. Differentiation of both leaf and roots occurred on medium having 2 per cent sucrose (Table 5.5). Thus, amongst all the carbohydrate sources both leaf and root differentiation occurred only in seedlings developing on medium having 2 per cent sucrose.

Whereas the seedlings on medium incorporated with 1 per cent glucose/ fructose were pale green; those growing on 2 per cent glucose/ fructose supplemented medium were light green. Moderately green seedlings developed on medium having 3 per cent glucose/ fructose and 1 and 4 per cent sucrose. Seedlings grown on medium having 2–3 per cent sucrose were darker green in colour. The seedlings which developed on medium containing 4 per cent glucose/fructose remained white. Thus amongst the various carbohydrate sources used in the present study better development of pigments occurred in the presence of 2–3 per cent sucrose (Table 5.5).

Nitrogen sources: Amongst nitrogenous sources (L–arginine, L–aspartic acid, L–phenylalanine, glycine, ammonium nitrate, sodium nitrate, calcium nitrate, potassium nitrate and ammonium sulphate) used, germination of seeds and seedling differentiation occurred only on medium supplemented with $NH_4 NO_3$ and $NH_4 SO_4$ (Table 5.6). No seed germination occurred on medium supplemented with organic nitrogen sources, L–arginine, L–aspartic acid, L–phenylalanine and glycine (Table 5.6).

Control: *Paphiopedilum* seeds did not germinate on medium devoid of any nitrogen sources (Table 5.6, Fig. 5.10).

Treated: Germination of seeds occurred by the incorporation of 0.1–1.0 gl^{-1} of $NH_4 NO_3/NH_4 SO_4$ into the medium. No seeds germinated on the medium having 2.0 g 1^{-1} of $NH_4 NO_3/NH_4 SO_4$. Seeds germinated on the 25th day on medium having 0.1–1.0 g 1^{-1} $NH_4 NO_3/NH_4 SO_4$ (Table 5.6). Germination of seeds increased with the increasing concentration of $NH_4 NO_3$ /$NH_4 SO_4$ (Fig. 5.10). Maximum seed germination occurred when 0.5 g 1^{-1} of $NH_4 NO_3/NH_4 SO_4$ were incorporated in the medium (Fig. 5.10). With a further increase in their concentration seed germination declined (Fig. 5.10). Suitability of nitrogen source was thus in the following order: $NH_4 NO_3$ and then $NH_4 SO_4$.

The seedlings developing on medium incorporated with $NH_4 NO_3$ differentiated leaf. Differentiation of both leaf and roots occurred in the seedlings growing on medium having 0.5 g 1^{-1} $NH_4 NO_3$. But seedlings developing on $NH_4 SO_4$ supplemented medium did not differentiate beyond the protocorm stage (Table 5.6).

The seedlings growing on $NH_4 NO_3$ incorporated medium were moderately green while seedlings developed on $NH_4 SO_4$ containing medium remained white (Table 5.6).

Growth hormones: The effect of growth hormones on seed germination and subsequent growth is given in Table 5.7.

Table 5.6 : Effect of different concentrations of various inorganic nitrogen source on seed germination and seedling growth in *P. spicerianum*

Character	*Control*	*Inorganic nitrogens (g ml^{-1})*																			
		Ammonium nitrate				*Sodium nitrate*				*Calcium nitrate*				*Potassium nitrate*				*Ammonium sulphate*			
		0.1	*0.5*	*1.0*	*2.0*	*0.1*	*0.5*	*1.0*	*2.0*	*0.1*	*0.5*	*1.0*	*2.0*	*0.1*	*0.5*	*1.0*	*2.0*	*0.1*	*0.5*	*1.0*	*2.0*
Time taken to commence germination (days)	–	25	25	25	–	–	–	–	–	–	–	–	–	–	–	–	–	25	25	25	–
Germination per cent	0	48	70	52	0	0	0	0	0	0	0	0	0	0	0	0	0	39	61.5	41	0
Differentiation after 90 days	–	++	+++	++	–	–	-	-	-	–	–	–	-	–	–	–	-	+	+	+	–
Pigmentation of seedlings	–	MG	MG	MG	–	–	–	–	–	–	–	–	-	-	–	–	–	W	W	W	–

\+ protocorm stage

++ seedlings exhibiting leaf differentiation

+++ seedlings exhibiting leaf and root differentiation

MG moderately green

W white

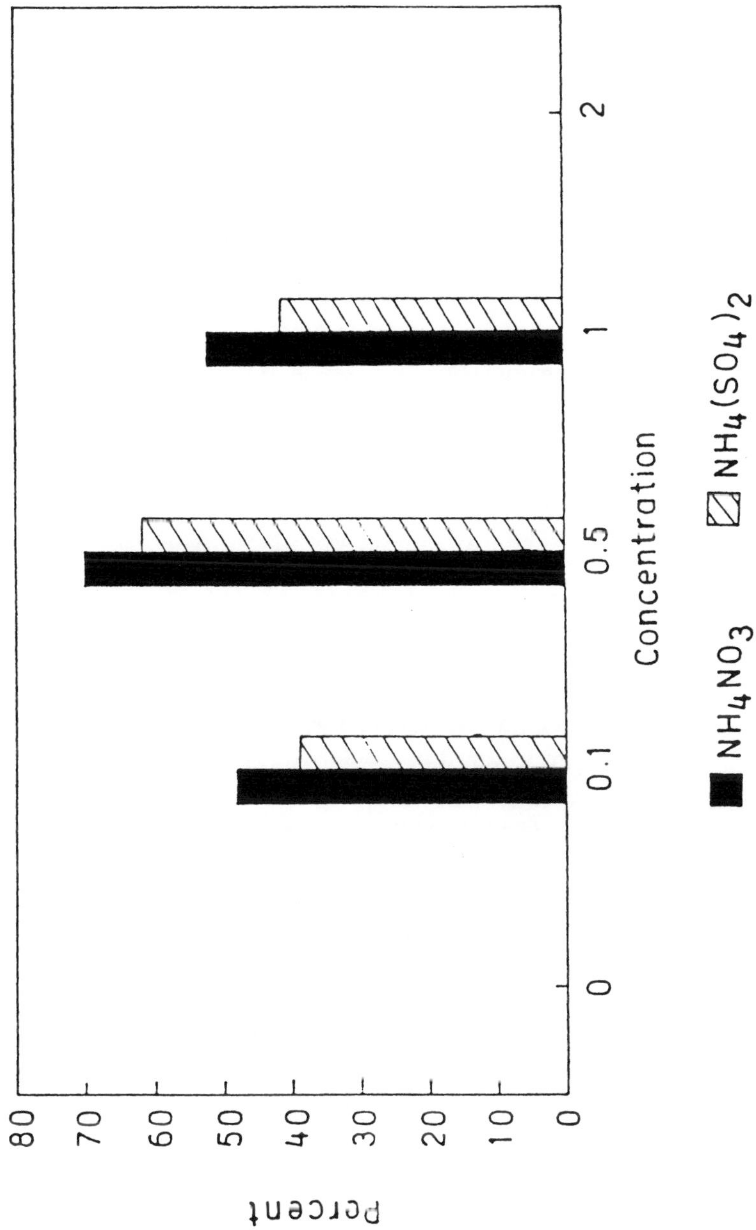

Fig. 5.10 : Inorganic nitrogen effect on P. *spicerianum* seed germination

Control: On medium devoid of growth hormones, germination of *Paphiopedilum* seeds commenced on the 25th day; and altogether 78.5% seeds germinated (Table 5.7, Fig. 5.11). 90 days old seedlings differentiated both leaves and roots. The differentiated seedlings were darker green in colour (Table 5.7).

Treated: Whereas 0.1–0.5 mg 1^{-1} kinetin in the germinating medium did not influence commencement of seed germination, the seed germination was delayed when 1.0 mg 1^{-1} kinetin was used. No seed germination occurred at 5.0–10.0 mg 1^{-1} kinetin concentration. Incorporation of BAP (0.1–0.5 mg 1^{-1}) delayed commencement of seed germination. No seed germination occurred in presence of 1.0–10.0 mg 1^{-1} BAP. 0.1 mg 1^{-1} NAA though reduced the time required for seed germination addition of 0.5–5.0 mg 1^{-1} NAA in the medium had no influence on the commencement of seed germination. No seeds germinated at 10.0 mg 1^{-1} NAA. Whereas IAA concentrations 0.1–0.5 mg 1^{-1} reduced the time required for seed germination, the IAA 1.0–5.0 mg 1^{-1} had no influence on the commencement of seed germination. No germination of seeds occurred at 10.0 mg 1^{-1} IAA (Table 5.7). On medium supplemented with 0.1–0.5 mg 1^{-1} 2, 4–D seed germination commenced at the same time as control, but no seeds germinated when 1.0–10.0 mg 1^{-1} 2, 4–D was incorporated in the medium. 0.1 mg 1^{-1} GA_3 did not alter the commencement of seed germination; however, seed germination was delayed when 0.5–5.0 mg 1^{-1} GA_3 was used. Seeds did not germinate on medium containing 10.0 mg 1^{-1} GA_3.

Kinetin; BAP; 2, 4–D and GA_3 inhibited seed germination and the inhibition increased with the increasing concentration (Table 5.7, Fig. 5.11, 5.12). Conversely, NAA (0.1–0.5 mg 1^{-1}) and IAA (0.1–1.0 mg 1^{-1}) enhanced seed germination over control (Fig. 5.12). But germination of seeds decreased with the increasing concentrations of NAA and IAA in the medium (Table 5.7, Fig. 5.12). Compared to NAA, IAA was more effective in stimulating seed germination.

Seedlings developing on medium combining kinetin (0.1 mg 1^{-1}) differentiated both leaf and root but only leaves differentiated in seedlings developing on 0.5–1.0 mg 1^{-1} kinetin supplemented medium. No root differentiation occurred at these concentrations of kinetin. No differentiation beyond the protocorm stage occurred in the presence of BAP, NAA, IAA and 2, 4–D. Whereas in the presence of 0.1–0.5 mg 1^{-1} GA_3 seedlings differentiated only leaves no differentiation beyond protocorm stage occurred when 1.0–5.0 mg 1^{-1} GA_3 was incorporated in the medium (Table 5.7).

Growth hormones influence pigmentation of seedlings 0.1–0.5 mg 1^{-1} kinetin and GA_3 concentrations resulted in moderately green seedlings but light green seedlings differentiated on medium having BAP and 1.0 mg 1^{-1} kinetin. The seedlings did not develop green colour in presence of 1.0–5.0 mg 1^{-1} GA_3 and NAA, IAA and 2, 4–D concentrations (Table 5.7).

Table 5.7: Effect of cifferent concentrations of various growth hormones on seed germination, seedling differentiation and pigmentation of the seedlings in *P. spicerianum.*

Character	*Control*	*Growth hormones (mg 1-)*																													
		KN					BAP					NAA					IAA					2, 4-D					GA_3				
		0.1	0.5	1.0	5.0	10.0	0.1	0.5	1.0	5.0	10.0	0.1	0.5	1.0	5.0	10.0	0.1	0.5	1.0	5.0	10.0	0.1	0.5	1.0	5.0	10.0	0.1	0.5	1.0	5.0	10.0
Time taken to commence germination (days)	25	25	25	30	-	-	30	30	-	-	-	20	25	25	25	-	20	20	25	25	-	25	25	-	-	-	25	35	35	35	-
Germination per cent	78.5	69	65	48	0	0	56	18	0	0	0	88	80	70	18	0	92	90	85	54	0	49	9	0	0	0	55	49	40	17.5	0
Differentiation ater 90 days	+++	+++	++	++	-	-	+	+	-	-	-	+	+	+	+	-	+	+	+	+	-	+	+	-	-	-	++	++	+	+	-
Pigmentation of seedlings	DG	MG	MG	LG	-	-	LG	LG	-	-	-	W	W	W	B	-	W	W	W	Br	-	PG	Br	-	-	-	MG	MG	W	W	-

+	protocorm stage	LG	light green
++	seedlings exhibiting leaf differentiation	DG	dark green
+++	seedlings exhibiting leaf and root differentiation	Br	brown
PG	pale green	B	black
MG	moderately green	W	white

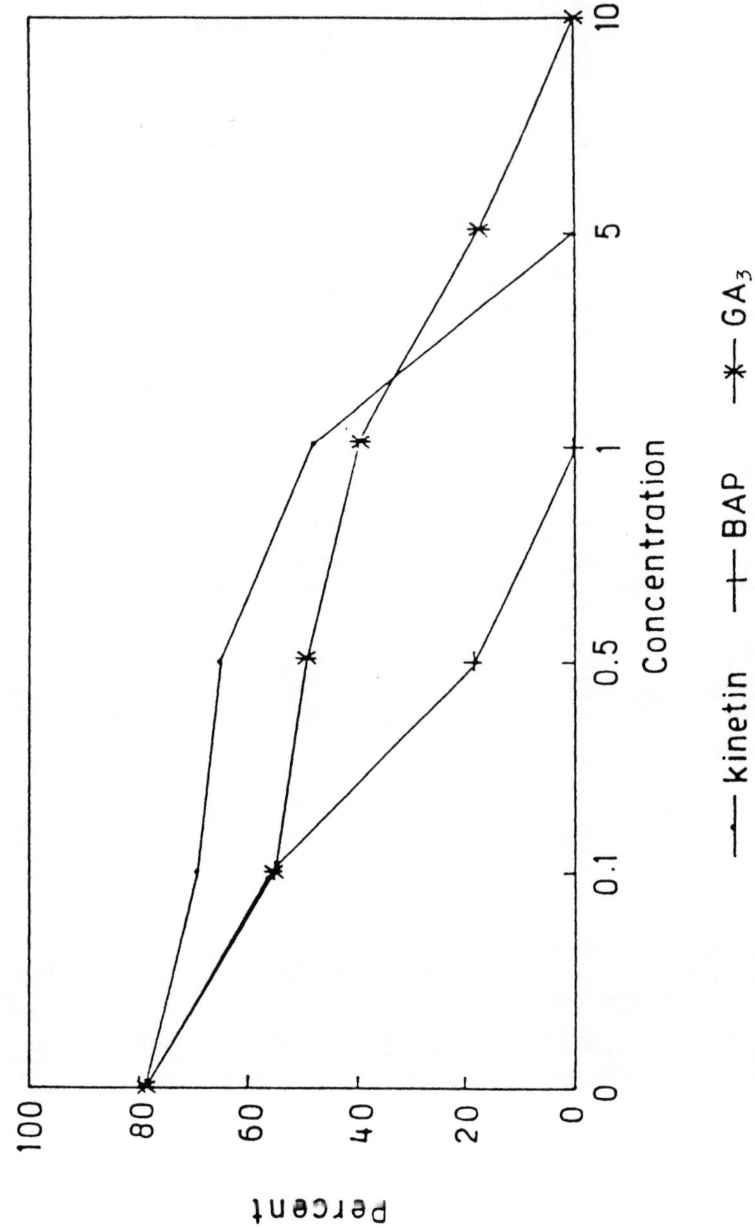

Fig. 5.11 : Cytokinin & GA_3 effect on *P. spicerianum* seed germination

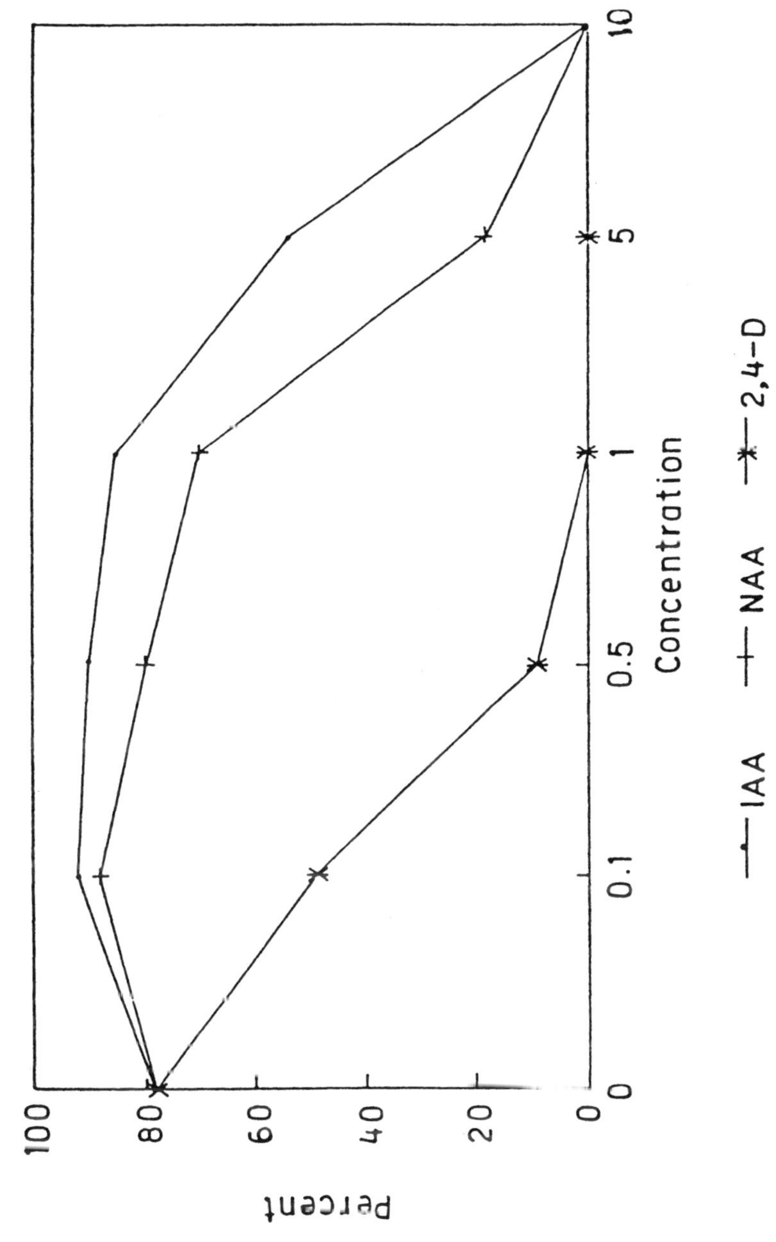

Fig. 5.12 : Auxin effect on *P. spicerianum* seed germination

Table 5.8: Effect of various organic supplements on seed germination and seedling grrowth in *P. spicerianum.*

Character	*Control*	*Organic supplements*				
		Coconut Milk	*Banana Homogenate*	*Apple Juice*	*Peptone*	*Activated Charcoal*
Time taken to commence germination (days)	25	25	—	—	30	25
Germination per cent	78.5	68.0	0.0	0.0	58.0	65.0
Differentiation after 90 days	+++	+++	—	—	++	+++
Pigmentation of seedlings	DG	DG	—	—	MG	DG

+ protocorm stage
++ seedlings exhibiting leaf differentiation
+++ seedlings exhibiting leaf and root differentiation
DG dark green
MG moderately green

Organic supplements: The effect of organic supplements on seed germination is given in Table 5.8.

Control: On medium devoid of any organic supplements, *Paphiopedilum* seeds germinated on the 25th day. The degree of germination was 78.5%. 90 days old seedlings differentiated both leaves and roots and were darker green in colour (Table 5.8).

Treated: Commencement of seed germination was not affected by CM/activated charcoal but peptone delayed germination. No seeds germinated on medium having banana homogenate/apple juice (Table 5.8). Organic supplements inhibited seed germination and the decrease was in the following order: CM > activated charcoal > peptone.

The seedlings developing on medium supplemented with either CM or activated charcoal exhibited differentiation of both leaf and roots but only leaf differentiation occurred when peptone was used (Table 5.8).

Seedlings growing on medium having either CM or activated charcoal were darker green while peptone developed moderately green seedlings (Table 5.8).

DISCUSSION

Carbohydrate sources: In the present investigation carbohydrates play an important role in seed germination and seedling growth of both the genera. Seeds of *D. chrysanthum* exhibited very poor seed germination on medium devoid of carbohydrate sources. The differentiation beyond the protocorm stage also did not occur. In *Cymbidium* CYGNUS "Inghams" also seeds do not develop beyond protocorm stage (Fonnesbech, 1972). Seeds of *P. spicerianum,* did not germinate at all on medium lacking carbohydrate source. Thus incorporation of a carbohydrate source in the germinating medium is essential for germinaton of seeds and differentiation beyond protocorm stage. This requirement of carbohydrate could be due to lack of endosperm,the tissue which nourishes developing embryo, in the orchid seeds.

In both *D. chrysanthum* and *P. spicerianum,* all the carbohydrates with the exception of lactose and soluble starch proved suitable as sources of carbohydrates for seed germination and subsequent growth. Cell membranes which may be impermeable to starch (Arditti and Ernst, 1984; Ernst *et al.*, 1971; Hew and Mah, 1989) could be permeable to disaccharides (Ernst and Arditti, 1990) and monosaccharides (Ernst *et al.*, 1971). Amongst disaccharides lactose revealed a comparatively poor germination and subsequent differentiation in *D. chrysanthum*. *P. spicerianum* seeds, however, did not germinate on lactose containing medium. Lactose containing D-galactopyranosyl moiety is considered responsible for poor performance in *Phalaenopsis* (Ernst, 1967; Ernst and Arditti, 1990) and *Dendrobium* (Ernst, 1967). Ernst (1967) reported that with the sole exception of D-galactose all D-hexose sugars are suitable as sources of carbon. Ernst (1967) reported death of seedlings exposed to galactose. Ernst (1967) therefore concluded that relatively poor performance of the disaccharide, lactose containing the D-galactopyranosyl moiety, are not entirely unexpected.

For seed germination and subsequent differentiation of both *D. chrysanthum* and *P. spicerianum,* sucrose (2%) was suitable. Higher concentrations of glucose, fructose, sucrose, lactose in case of *D. chrysanthum* and higher concentrations of glucose, fructose, and sucrose in case of *P. spicerianum* were detrimental for seed germination and subsequent seedling growth. The detrimental effect is possibly due to adverse effect of higher concentrations of carbohydrates on the osmoregulation of the tissues. Similar, were the findings of Mathews and Rao (1985) on *Vanda* hybrid.

Nitrogen sources: In the present investigation, in the absence of nitrogen, germination of *D. chrysanthum* seeds was poor, and the germinated seeds did not differentiate beyond the protocorm stage. No germination of *P. spicerianum,* seeds occurred in absence of nitrogen. Unlike organic nitrogen (L-arginine, L-aspartic acid, L-phenylalanine, glycine) inorganic nitrogen, when incorporated in the basal medium, induced seed germination and growth.

Thus for inducing germination of seeds and growth, whereas nitrate salts were effective in *D. chrysanthum,* the ammonium salts were effective in *P. spicerianum*. Burgeff (1936) also stated that nitrate as nitrogenous source was best used by epiphytic orchid species, but those species which germinate in dark require specifically ammonium salts as nitrogenous source. Unlike *P. spicerianum* in the present investigation seeds of *D. chrysanthum,* eventhough germinated best when nitrate was used as nitrogen source, these also germinated on medium containing ammonium as nitrogen source. Withner (1959) also found that *Paphiopedilum* unlike other genera of orchids is very specific in its nitrogen requirement. It may therefore be concluded that best germination of *D. chrysanthum* seeds can be achieved by using nitrate as nitrogen source. *P. spicerianum* seeds on the other hand would specifically require ammonium as the nitrogen source.

The organic nitrogen sources viz. glycine, phenylalanine, L-arginine, L-aspartic acid as nitrogen supplements in the medium could not induce seed germination in both *D. chrysanthum* and *P. spicerianum*. Inhibitory effect of glycine (Raghavan, 1964; Raghavan and Torrey, 1964), phenylalanine and aspartic acid (Arditti, 1967; Arditti and Ernst, 1984) have been reported in literature. Arginine has been reported to replace ammonium nitrate requirement in *Cattleya* (Arditti and Ernst, 1984). However, in present investigation arginine could not substitute ammonium nitrate requirements for seed germination in both the genera investigated.

In the present investigation whereas calcium nitrate caused maximum seed germination and development of well differentiated (root, shoot, leaf) chlorophyllous seedlings in *D. chrysanthum*. Ammonium sulphate was most suitable for seed germination and subsequent seedling growth of *P. spicerianum*. Thus, it can be concluded that, unlike organic nitrogen source, inorganic nitrogen sources are suitable for seed germination and subsequent seedling growth in both the genera. Amongst inorganic nitrogen sources, calcium nitrate was suitable for *D.chrysanthum* while ammonium sulphate was best for *P. spicerianum* seed germination and subsequent seedling growth.

Growth hormones: In the present study, in *D. chrysanthum* kinetin eventhough did not improve germination of seeds it induced better shoot and root growth. BAP supplemented medium, on the other hand caused unsatisfactory germination and growth vis-a-vis control and kinetin supplemented medium. Auxins inhibited germination and growth of *D. chrysanthum*. These findings are in agreement with Kano (1965), Miyazaki and Nagamatsu (1965). 2, 4-D in *D. chrysanthum* developed abnormal growth. Sun (1955) stated that 2, 4-D apparently disrupts the orderly metabolic and physiological processes, and thus leading to non polarized cell division and abnormal growth pattern. 2, 4-D also disturbed the chlorophyll synthesis. 2, 4-D caused disturbance of chlorophyll synthesis has been reported as the reason for abnormal growth (Fonnesbech, 1972) in *Cymbidium* CYGNUS " Inghams".

In *P. spicerianum* cytokinins had no beneficial effect. However, auxin supplemented medium with the exception of 2, 4-D revealed a speedy and increased seed germination compared to control. These results corroborate the findings of Arditti and Ernst (1984). However, further growth of the germinated seeds of *P. spicerianum* was inhibited in auxin supplemented medium. GA_3 has an inhibitory effect on seed germination in both the genera (*D. chrysanthum* and *P. spicerianum*). With the incorporation of GA_3 in the medium though subsequent seedling growth remained unaffected in *D. chrysanthum* it was inhibited in *P. spicerianum*. These findings are in agreement with Arditti and Ernst (1984) who observed that in general, the effects of exogenous gibberellins on the growth of orchid seedling are usually negative. Thus, it can conclude that auxin, with the exception of 2, 4-D, induce speedy and increased seed germination in *P. spicerianum* but inhibit seed germination in *D. chrysanthum*. However, in both the genera subsequent growth of the germinated seeds on the auxin containing medium was inhibited. Kinetin induced better shoot and root growth in *D. chrysanthum*, but in *P. spicerianum*, kinetin had no beneficial effect.

Organic supplements: Organic supplements inhibited germination of seeds in both the genera. However, seedling differentiation and development of pigments remained unaffected with CM and activated charcoal. *D. chrysanthum* exhibited poor growth on banana homogenate, apple juice and peptone supplemented medium. But *P. spicerianum* revealed no germination on banana homogenate and apple juice medium and exhibited poor growth on peptone supplemented medium. The present investigation therefore, suggest that both *D. chrysanthum* and *P. spicerianum* do not require the incorporation of organic supplements for seed germination and subsequent differentiation.

6
Meristem Culture

Introduction

To avoid segregation due to heterozygosity asexual propagation is the only means to obtain a large number of selected recombinant types in the outbreeding taxa of Orchidaceae (Wilfret, 1966). Vegetative propagation in orchids is a very slow process (Singh, 1976). Subsequent to the discovery of totipotency of plant cells (Haberlandt, 1902) and role of auxin and cytokinin in shoot and root initiation (Skoog and Miller, 1957), mass clonal propagation of various species have been attempted using tissue culture techniques (Murashige, 1974, 1977; Withner, 1974; Arditti, 1977). Orchid *Cymbidium* is the first horticultural plant species successfully mass propagated using tissue culture technique, (Morel, 1960). The application of tissue culture technique, for rapid clonal propagation, has been highly successful in a few orchids (Rao, 1977; Hughes, 1981). Since Morel's work in 1960, shoot tips remained the most commonly used explants (Wimber, 1963; Wilfret, 1966; Steward and Mapes, 1971; Werckmeister, 1971 and Fonnesbech, 1972). While Morel (1960) used very small explants (0.1 mm thick), other workers used much larger explants (about 3 to 5 mm thick). Shoot tips of *Dendrobium* when cultured also developed protocorm-like bodies after one month (Kim *et al.*, 1970; Morel, 1974). Other plant parts like roots of *Phalaenopsis, Dendrobium;* apical shoot/axillary buds of *Cymbidium, Dendrobium , Oncidium, Vanda* and leaves of *Cymbidium, Dendrobium* have also been used (Sagawa and Kunisaki, 1984). However, in orchids scarcity of explant sources are usually the limiting factors for clonal propagation of orchids using tissue culture technique. Sagawa and Kunisaki (1984) suggested that besides shoot tips and axillary buds inflorescences, leaves and roots can be used as explant sources thus overcoming scarcity of explants in orchids. But for *Paphiopedilum* and *Phalaenopsis* no commercially feasible method for rapid clonal propagation has been developed yet.

Therefore, in the present investigation *in vitro* clonal multiplication of *Dendrobium chrysanthum* and *Paphiopedilum spicerianum* has been attempted by obtaining explants from *in vivo* and *in vitro* grown seedlings of these species.

MATERIALS AND METHODS

Dendrobium and *Paphiopedilum* explants for culturing were obtained as detailed below:

Explants were obtained from seedlings raised *in vitro* (1.0–1.5 cm long) and those collected (4.0–5.0 cm long) from nature for axillary bud and inflorescence culture. Root tips and shoot tips (after removing leaves around the shoot apex) measuring 3 to 4 mm were excised. Leaf segments measuring 3 to 4 mm were sectioned from leaves developing on *in vivo* and *in vitro* grown seedlings. Besides, leaves measuring 4 to 5 mm were also taken from *in vitro* raised seedlings. Adult plants occurring in nature and bearing axillary buds were used for obtaining axillary bud explants. *D. chrysanthum* explants were also gathered from the adult plants having inflorescence primordia. Stems bearing axillary buds (3 mm × 3 mm) were collected. The leaves were removed by stripping and the axillary buds were excised from stem by making two incisions at right angles to each other. Stems of *D. chrysanthum* bearing inflorescence primordia were collected. The inflorescence primordia measuring 2 mm × 2 mm, 2 mm × 3 mm, 3 mm × 3 mm and 4 mm × 3 mm were excised from the stem like axillary buds.

Surface sterilization of the explants obtained from *in vitro* grown seedlings was not done. Explants from *in vivo* plants were surface sterilized for 20 minutes by immersion in 10 per cent sodium hypochlorite solution and rinsing three times in sterile distilled water under aseptic conditions. The explants of *P. spicerianum*, however, were again surface sterilized with 0.01 per cent mercuric chloride for 30 seconds by constant agitation and rinsing thrice with sterile distilled water. The explants were cultured on appropriate semi-solid culture medium with their cut ends, in case of shoot tip, axillary bud and inflorescence primordia, inserted in the medium. The explants of leaf parts and root tips were, however, placed on medium. Per container only one explant was used and 15 replicates were maintained for each treatment.

Preparation of Culture Medium

Thomale GD (Th GD) (1954) medium, Vacin and Went (VW) (1949) medium, Knudson C (Kn C) (1946) medium and Knudson C medium with various combinations of 6–furfurylaminopurine (Kin); benzyladenine (BA); α– naphthyl acetic acid (NAA); 3–indole acetic acid (IAA); 2, 4-dichlorophenoxy acetic acid; coconut milk (CM) 15 per cent and coconut milk+ activated charcoal 0.1 per cent (CM + AC) were used for shoot induction. Kn C medium devoid of growth hormones served as control.

Kn C medium supplemented with 50 g 1^{-1} banana homogenate (preparation as in chapter 5) was used to develop roots in the shoots (15 to 20 mm long) developed from *in vitro* cultures. The nutrient media were gelled with 0.9 per cent bacto agar and pH of the medium was adjusted to 5.5. Twenty ml of medium was dispensed into corning (25 mm × 200 mm) test tubes and conical flasks (100 ml). The test tubes having medium were plugged with non-absorbent cotton plugs and autoclaved at 1 Kg cm^{-2} (121°C) for 20 minutes.

Culture Conditions

The cultures were incubated upto 180 days at 25 ± 2°C and 16 hr, 3000 lux fluorescent light.

Data Collection

1. Shoot induction in various explants (shoot tip, root tip, leaf parts, axillary buds and inflorescence primordia) on Th GD, VW, Kn C and variously modified Kn C media was scrutinised on 60, 90 and 180 days.
2. In cultured primoridal inflorescences, besides the induction of shoot, shoot length, leaf number, leaf length and leaf breadth were recorded after 180 days.

RESULTS

Effect of Media: None of the media (Kn C, Th GD and VW) supported shoot induction on the cultured explants upto 180 days. However, shoot tips and root tips from *in vitro* seedlings, and axillary bud and inflorescence primordia (3 mm × 3 mm) from *in vivo* plantlets of *D. chrysanthum* remained green on Kn C medium. All the other explants of *D. chrysanthum*, turned white including those of *P. spicerianum*. Explants cultured on Kn C medium supplemented with various concentrations (0.01–10.0 mg 1^{-1}) of NAA, IAA, 2, 4–D and kinetin either individually or in combination also did not develop shoots from the explants of both the genus (Table 6.1). However, in *D. chrysanthum* shoot tips differentiated shoots on Kn C medium incorporated with CM, while root tips differentiated shoots on Kn C medium incorporated with CM and activated charcoal. Axillary buds differentiated on both Kn C medium + CM and Kn C medium + CM + activated charcoal. Inflorescence primordia, however, differentiated shoots only on Kn C medium incorporated with BA (0.01 to 1.0 mg 1^{-1}).

Table 6.1 : Effect of NAA, IAA, 2, 4–D and Kinetin on shoot development from various explants of both *D. chrysanthum* and *P. spicerianum*.

		KINETIN (mg 1⁻¹)					
		0.0	*0.01*	*0.1*	*0.5*	*1.0*	*10.0*
NAA	0.0	–	–	–	–	–	–
	0.01	–	–	–	–	–	–
	0.1	–	–	–	–	–	–
	0.5	–	–	–	–	–	–
	1.0	–	–	–	–	–	–
	10.0	–	–	–	–	–	–
IAA	0.0	–	–	–	–	–	–
	0.01	–	–	–	–	–	–
	0.1	–	–	–	–	–	–
	0.5	–	–	–	–	–	–
	1.0	–	–	–	–	–	–
	10.0	–	–	–	–	–	–
2, 4 –D	0.0	–	–	–	–	–	–
	0.01	–	–	–	–	–	–
	0.1	–	–	–	–	–	–
	0.5	–	–	–	–	–	–
	1.0	–	–	–	–	–	–
	10.0	–	–	–	–	–	–

Shoot tip explants obtained from *in vitro* raised seedlings of *D. chrysanthum*, after 60 days of culturing on Kn C medium supplemented with CM, developed 9–11 protocorm-like bodies (Plate 6a.1) from the base of the explant (Table 6.2, Plate 6a.1). These in 90 days (Table 6.2) differentiated shoots (Plate 6a.2) and in 180 days the shoots became two leaved and measured 4-5 mm long. Shoot induction was evident only in 26.6 per cent explants (Table 6.3). Shoot tip explants obtained from *in vivo* grown seedlings of *D. chrysanthum*, however, did not form

Table 6.2 : The effect of Kn C medium incorporated with different growth regulators on *in vitro* shoot production from *D. chrysanthum* explants.

Medium	*In vitro explants*												*In vivo Explants*											
	Shoot tips						*Root tips*						*Axillary buds*						*Inflorescence primordia*					
	Time taken in days			*Numbers/Explant*			*Time taken in days*			*Numbers/ Explant*			*Time taken in days*			*Numbers/ Explant*			*Time taken in days*			*Numbers/ Explant*		
	Callus Deve-lop-ment	*Plb deve-lop-ment*	*Shoot deve-lop-ment*	*Root deve-lop-ment*	*Plb*	*Shoot*	*Callus Deve-lop-ment*	*Plb deve-lop-ment*	*Shoot deve-lop-ment*	*Root deve-lop-ment*	*Plb*	*Shoot*	*Callus Deve-lop-ment*	*Plb deve-lop-ment*	*Shoot deve-lop-ment*	*Root deve-lop-ment*	*Plb*	*Shoot*	*Callus Deve-lop-ment*	*Plb deve-lop-ment*	*Shoot deve-lop-ment*	*Root deve-lop-ment*	*Plb*	*Shoot*
KnC + CM	-	60	90	-	9 to 11	9 to 11	-	-	-	-	-	-	-	-	60	-	-	1	-	-	-	-	-	-
KnC + CM + AC	-	-	-	-	-	-	60	90	110	-	-	9 to11	-	-	60	-	-	1	-	-	-	-	-	-
KnC + BA 0.01 mgl⁻¹	-	-	-	-	-	-	-	-	-	-	-	-	-	-	-	-	-	-	-	-	90	-	-	1
+ 0.1 mgl⁻¹	-	-	-	-	-	-	-	-	-	-	-	-	-	-	-	-	-	-	-	-	60	-	-	1 to 6
+ 0.5 mgl⁻¹	-	-	-	-	-	-	-	-	-	-	-	-	-	-	-	-	-	-	-	-	60	-	-	1 to 2
+ 1.0 mgl⁻¹	-	-	-	-	-	-	-	-	-	-	-	-	-	-	-	-	-	-	-	-	90	-	-	1

Table 6.3 : Percentage of response to shoot induction by *in vitro* cultured shoot tip, root tip and axillary bud explants after 180 days

	Shoot tips			*Root tips*			*Axillary buds*		
Medium	*Explants cultured*	*Explants forming shoots*	*Percentage of shoot induction*	*Explants cultured*	*Explants forming shoots*	*Percentage of shoot induction*	*Explants cultured*	*Explants forming shoots*	*Percentage of shoot induction*
KnC + CM	15	4	26.6	15	0	0	15	7	46.6
KnC + CM + AC	15	0	0	15	3	20.0	15	8	53.5

Plate 1 a: Flowers of *Dendrobium* species

1. *D. aphyllum*
2. *D. chrysanthum*
3. *D. chrysotoxum*
4. *D. falconeri*
5. *D. heterocarpum*
6. *D. ochreatum*
7. *D. primulianum*
8. *D. wardianum*

1
4
5
6
3
8

Plate 6b: Shoot formation in inflorescence primordia explants of *D. chrysanthum*, 180 days after culture

1. inflorescence primordia at the time of inoculation
2. multiple shoot formation on Kn C medium containing 0.1 mg 1^{-1} BA
3. abnormal single shoot with rosette of leaves on Kn C medium containing 0.1 mg 1^{-1} BA
4. differentiation of single vegetative shoot and floral bud from the same explant on Kn C medium containing 1.0 mg 1^{-1} BA
5. floral buds differentiated on Kn C medium containing 5.0-10.0 mg 1^{-1} BA
6. multiple shoot formation on Kn C medium containing 0.1 mg 1^{-1} BA and 0.01 mg 1^{-1} NAA
7. single shoot formation on Kn C medium containing 0.1 mg 1^{-1} BA and 0.05 mg 1^{-1} NAA
8. explants cultured on Kn C medium containing 0.1 mg 1^{-1} BA and 0.1 mg 1^{-1} NAA
9. rooted plantlets from sub-cultured shoots (10.0-15.0 mm), 60 days of subculturing

Plate 7b.

shoots on any of the medium tested. In case of *P. spicerianum* no shoot induction occurred in shoot tip explants obtained from *in vitro* grown seedlings. The cultures of shoot tip explants of *P. spicerianum* obtained from *in vivo* grown seedlings got contaminated most of the time and thus their response could not be investigated.

Callusing occurred on the 60th day in root tip explants obtained from *in vitro* grown seedlings of *D. chrysanthum* and cultured on Kn C medium + 15 per cent CM + 0.1 per cent activated charcoal (Table 6.2). Differentiation of 9–11 protocorm like bodies per callus was evident on the 90th day of culturing (Table 6.2). The protocorm like bodies differentiated 4–5 mm long shoots in 180 days old cultures (*Plate 6a. 3*). Shoot induction was evident only in 20 per cent root tip explants (Table 6.3). In *D. chrysanthum* whereas only root tips obtained from *in vivo* grown seedlings revealed no callusing upto 180th day, the root tips obtained from both *in vivo* and *in vitro* grown seedlings of *P. spicerianum* exhibited no callusing on any of the nutrient medium up to the 180th day. Shoot induction was not observed in leaf explants obtained either from *in vitro* or *in vivo* grown plantlets of *D. chrysanthum* and *P. spicerianum.*

Axillary buds, 46.6 per cent and 53.3 per cent respectively, cultured on Kn C medium + CM and Kn C medium + CM + activated charcoal, on the 60th day of culturing, revealed development of single shoots (Table 6.2 & 6.3). Axillary buds of *P. spicerianum,* however, did not develop any shoots on any of the nutrient medium.

The cultured explants bearing (3.0 mm × 3.0 mm) inflorescence primordia (*Plate 6b.1*) differentiated shoots on Kn C medium supplemented with BA (0.01 to 1.0 mg l^{-1}) (*Plate 6b. 2–4*). On Kn C medium containing 0.1 and 0.5 mg l^{-1} BA 80 per cent and 40 per cent explants respectively differentiated shoots on the 60th day (Table 6.2 & 6.4). But on Kn C medium having 0.01 and 1.0 mg l^{-1} BA only 26.6 and 33.3 per cent explants respectively differentiated shoots on the 90th day (Table 6.2 & 6.4). The percentage of explants forming shoots thus increased with the increasing concentrations of BA in the medium up to 0.1 mg l^{-1} and thereafter decreased (Table 6.4). However, when BA concentration higher than 1.0 mg l^{-1} was used, instead of shoots, floral buds differentiated (*Plate 6b.5*). The optimum concentration of BA for shoot induction was 0.1 mg l^{-1} (Table 6.2 & 6.4). It induced differentiation of shoots in 80 per cent explants, of which 75 per cent developed multiple shoots (4–6) (*Plate 6b.2*) and 25 per cent developed abnormal single shoots having condensed, internodes and rossette of leaves (*Plate 6b.3*). At 0.5 mg l^{-1} BA, amongst the shoot forming explants 50 per cent developed multiple shoots and 50 per cent developed single shoots but 0.01 mg l^{-1} BA induced only single shoots (Table 6.4). However, when explants were cultured on medium having 1.0 mg l^{-1} BA both single vegetative shoot and floral bud differentiated on the same explant (*Plate 6b. 4*).

Compared to normal single shoots differentiated in various treatment the number of leaves was significantly higher in the abnormal shoots differentiated on medium having 0.1 mg l^{-1} BA (Table 6.5). The number of leaves did not differ in single shoots differentiating on medium supplemented with 0.01 and 0.5 mg l^{-1} BA but leaf number was significantly reduced in shoots differentiating at 1.0 mgl^{-1} BA supplemented medium (Table 6.5). The size of single shoot developed

Table 6.4 : Effect of various concentrations of BA on *D. chrysanthum* explants after 180 days

Treatment BA concentrations ($mg\,l^{-1}$)	*Explants cultured*	*Explants forming shoot*				*Percentage of shoot induction*	
		Single shoot	*Multiple shoot*	*Total*	*Percentage response of explants*	*Single*	*Multiple*
0	15	–	–	–	–	–	–
0.01	15	4	–	4	26.6	100	–
0.1	15	3	9	12	80.0	25	75
0.5	15	3	3	6	40.0	50	50
1.0	15	5	–	5	33.3	100	–
5.0	15	–	–	–	–	–	–
10.0	15	–	–	–	–	–	–

Table 6.5 : Effect of various concentrations of BA on *D. chrysanthum* explants after 180 days

Treatments	*Single shoot*				*Multiple shoot*				
BA concen-trations ($mg1^{-1}$)	*Shoot height (mm)*	*Leaf number*	*Leaf length (mm)*	*Leaf breadth (mm)*	*Shoot number*	*Shoot height (mm)*	*Leaf number*	*Leaf length (mm)*	*Leaf breadth (mm)*
0.0	–	–	–	–	–	–	–	–	–
0.01	5.9 ± 0.3	4.0 ± 0.0	5.8 ± 0.5	2.0 ± 0.0	–	–	–	–	–
0.1	12.0 ± 1.1^{a}	6.7 ± 0.3^{a}	8.1 ± 0.6^{a}	2.0 ± 0.2	5.1 ± 0.3	4.2 ± 0.1	19.5 ± 0.4	11.0 ± 0.3	2.3 ± 0.1
0.5	15.3 ± 0.3^{ab}	4.0 ± 0.0^{b}	7.4 ± 0.0^{a}	2.0 ± 0.2	2.0 ± 0.0^{b}	4.0 ± 0.3	15.5 ± 0.3^{b}	7.7 ± 0.2^{b}	2.2 ± 0.1
1.0	8.0 ± 0.3^{abc}	2.4 ± 0.2^{abc}	4.3 ± 0.0^{abc}	1.3 ± 0.1^{abc}	–	–	–	–	–
5.0	–	–	–	–	–	–	–	–	–
10.0	–	–	–	–	–	–	–	–	–

1. ± standard error
2. a significantly different from 0.01 $mg1^{-1}$ BA
   b significantly different from 0.1 $mg1^{-1}$ BA
   c significantly different from 0.5 $mg1^{-1}$ BA
3. Students t test at $p = 0.05$

on explants cultured on medium supplemented with 0.5 mg l^{-1} was significantly higher than single shoots or other treatments (Table 6.5). The length of the leaf in the shoots developed on 0.1 mg l^{-1} and 0.5 mg l^{-1} BA supplemented medium did not differ and were significantly higher than those in single shoots formed at 0.01 and 1.0 mg l^{-1} BA (Table 6.5). Leaf breadth although did not differ at 0.01, 0.1 and 0.5 mg l^{-1} BA, it declined significantly at 1.0 mg l^{-1} BA (Table 6.5).

In comparison to the multiple shoots developed at 0.5 mg l^{-1} BA, the number of shoots, leaf number and leaf length were significantly higher in the multiple shoots developing from explants, cultured on Kn C medium supplemented with 0.1 mg l^{-1} BA (Table 6.5). But, shoot height and leaf breadth did not differ in the two treatments.

The explants cultured on medium supplemented with combinations of different concentration of IAA/2, 4–D (0.01–10.0 mg l^{-1}) and 0.1 mg l^{-1} BA failed to respond. But when combination of NAA (0.01 and 0.05 mg l^{-1}) and 0.1 mg l^{-1} BA was used the explants exhibited differentiation (Table 6.6, *Plate 6b.6 & 7*). However, this combination of NAA and BA 0.1 mg l^{-1} was inferior to the differentiation evident on medium supplemented with 0.1 mg l^{-1} of BA alone because relatively lesser number of explants developed shoots (Table 6.6). Further, the growth of shoot was also inferior in the combination treatment (Table 6.7). Besides, unlike BA treatment, in the combination treatment no abnormal single shoots differentiated (*Plate 6b.6 & 7*). Shoot development was completely inhibited (Table 6.7, *Plate 6b.8*), when higher NAA concentrations (0.1, 0.5 and 1.0 mg l^{-1}) were used to supplement Kn C medium having BA 0.1 mg l^{-1}.

Subculture of *in vitro* differentiated shoots (10-15 mm) on Kn C medium supplemented with 50 g l^{-1} banana homogenate alone resulted into rooted plantlets (*Plate 6b.9*).

DISCUSSION

During the present investigation Kn C medium proved better for the explants of *D. chrysanthum*. However, differentiation of shoots from various explants was better when Kn C medium was modified. On Kn C medium, supplemented with CM, shoot tip explants developed protocorm-like bodies which later produced shoots. Development of protocorm-like bodies from cultured shoot tips of *Dendrobium* and *Cymbidium* has been reported earlier also (Morel, 1965a, b; Sagawa and Kunisaki 1984; Kim *et al.*, 1970). In *Cymbidium* hybrids internodal areas between the embryonic leaves frequently expand and differentiate protocorm-like bodies (Wilfret, 1966). Addition of activated charcoal to the Kn C medium supplemented with CM induced callus formation on root tip explants of *in vitro* seedlings which later developed shoots. Because of the presence of inositol and ammonium nitrogen (Shantz and Steward, 1964) and growth regulators present in CM (Letham, 1974 and Van Staden & Drewes, 1975), the CM can support growth and differentiation of tissue. In *Cymbidium* the colour of the substrate may be of importance since roots tend to grow into black media (Werckmeister, 1970, 1971). Thus in the present investigation the activated charcoal induced stimulation of differentiation from roots could be due to its effect on the colour of

Table 6.6: Effect of various concentrations of NAA in combination with 0.1 $mg1^{-1}$ BA on shoot induction of *D. chrysanthum* explants after 180 days

Treatment NAA concentrations ($mg1^{-1}$)	*Explants cultured*	*Explants forming shoot* Single shoot	Multiple shoot	Total	Percentage response of explants	*Percentage of shoot induction* Single	Multiple
0.0	15	3	9	12	80	25	75
0.01	15	3	6	9	60	33	67
0.05	15	8	–	8	53.33	100	–
0.1	15	–	–	–	–	–	–
0.5	15	–	–	–	–	–	–
1.0	15	–	–	–	–	–	–

Table 6.7 : Effect of various concentrations of NAA in combination with 0.1 mg1^{-1} BA on *D. chrysanthum* explants after 180 days

Treatment	*Single shoot*				*Multiple shoot*				
NAA concentrations (mg 1^{-1})	*Shoot height (mm)*	*Leaf number*	*Leaf length (mm)*	*Leaf breadth (mm)*	*Shoot number*	*Leaf number*	*Shoot number (mm)*	*Leaf length (mm)*	*Leaf breadth (mm)*
0.0	12.0 ± 1.1	6.7 ± 0.3	8.1 ± 0.6	2.0 ± 0.2	5.1 ± 0.3	4.2 ± 0.1	19.5 ± 0.4	11.0 ± 0.3	2.3 ± 0.1
0.01	15.5 ± 0.5^a	5.0 ± 0.0	11.5 ± 0.1^a	2.3 ± 0.1	2.0 ± 0.0^a	3.3 ± 0.2	13.4 ± 1.0^a	8.1 ± 0.7^a	1.9 ± 0.1
0.05	15.0 ± 0.6^a	3.3 ± 0.2^{ab}	9.7 ± 0.2^{ab}	1.9 ± 0.0^b	–	–	–	–	–
0.1	–	–	–	–	–	–	–	–	–
0.5	–	–	–	–	–	–	–	–	–
1.0	–	–	–	–	–	–	–	–	–

± standard error

a significantly different from 0.0 mg 1^{-1} NAA

b significantly different from 0.01 mg1^{-1} NAA

students t test at $p = 0.05$

the culture medium. Root tip culture has been reported to be very useful, as multiplication of orchids would be possible without sacrificing either the main apex or lateral buds of mother plants (Rao, 1977). However, shoot induction was not evident on any medium in the shoot and root tip explants from *in vivo* plants. The differential shoot induction in the explants obtained from *in vitro* and *in vivo* grown seedlings could be due to differences in their physiological age. The importance of physiological age in regeneration has been emphasised by Vajrabhaya and Vajrabhaya (1970) in leaf culture studies of *Rhynchostylis gigantea*. On Kn C medium + CM, axillary bud explants from *in vivo* plants of *D. chrysanthum* differentiated directly into shoots. Suitability of CM for axillary bud has also been reported by Sagawa and Kunisaki (1984) in *Dendrobium* and *Cymbidium*. Kn C medium with BA (0.1 and 0.5 mg l^{-1}) proved suitable for shoot development in explants having inflorescence primordia and obtained from *in vivo* grown plants of *D. chrysanthum*. The essentiality of an exogenous supply of cytokinin for the induction of vegetative buds has been reported earlier as well (Tanaka *et al.*, 1988). Size of the inflorescence primordia explants was critical for shoot induction. Whereas inflorescence primordia 3 mm × 3 mm reverted to vegetative phase developed shoots, inflorescence primordia 4 mm × 3 mm differentiated floral buds.

Except axillary buds which revealed single shoots, multiple shoots developed from shoot tip, root tip and inflorescence primordia (3 mm × 3 mm) on suitable medium. Multiple shoots 9 to 11 differentiated from shoot tip and root tip explants and 4-6 from inflorescence primordia explants. However, percentage response of shoot tip, root tip explants to differentiation was less as compared to axillary bud and inflorescence primordia explants. On the 180th day, compared to shoots differentiated from shoot tip and root tip explants, the growth of shoots was better in the shoots which differentiated from inflorescence primordia and axillary bud explants. The growth of multiple shoots developed from explants cultured on 0.1 mg l^{-1} BA supplemented medium was the best. Thus in the present study inflorescence primordia were by far the best source of explants for successful clonal propagation followed by axillary buds, shoot tips and root tip explants.

However, none of the cultures developed roots in the respective medium. Shoots (1.0-1.5 cm) developed from inflorescence primordia, when subcultured on Kn C medium incorporated with banana homogenate, revealed root development. The effectiveness of banana in shoot and root development is believed to be due to cytokinins, gibberellins and auxins present in it (Steward, 1961). In case of *P. spicerianum* shoot induction was not evident on any medium in the various cultured explants obtained from *in vitro* and *in vivo* seedlings. Attempts to grow and propagate clonally *Paphiopedilum* hybrids through meristem culture have not succeeded well (Morel, 1974; Cooper, 1983; Huang, 1988). The apex of *Paphiopedilum* has very little protection from leaves. It is always contaminated by bacteria, and success is very limited as it is very difficult to get aseptic cultures (Morel, 1974). The present investigation supports the conclusions of Morel (1974).

7
Cold Preservation of Protocorms

Introduction

Conservation of bio-diversity is a fundamental requirement of plant breeding, horticulture and biotechnology (Wilkes, 1983; Holden and Williams, 1984). Selected genotypes of orchids, a highly heterozygous group, are maintained through vegetative propagation. However, maintaining a genotype in pots is not only risky but also expensive and time consuming (Bhojwani, 1981). Despite all the care taken, "some plants die from the effect of disease, virus and other unspecified causes" (Davies, 1970). *In vitro* storage is only one component of an alternative conservation strategy which also embraces propagation and movement of germplasm. Storage of cultures *in vitro* under normal growth conditions is costly and carries the risk of genetic change with time (Scowcroft *et al.*, 1985), contamination or accidental loss (Withers, 1985). Moreover, frequent sub-culturing of *in vitro* culture is required. Cold preservation is the most effective way for the conservation of germplasm *in vitro* (Galzy, 1969; Mullin and Schlegel, 1976; Tomes, 1979; Lundergan and Janick, 1979; Dale, 1980; Jacques *et al.*, 1982; Ng and Ng, 1991). Lowering of temperature reduces growth still maintaining a good survival rate (Ng and Ng, 1991). The advantages of these methods are: (a) relatively little space is needed for the preservation of large number of plants, (b) the plants are maintained free of pests, pathogens and virus, (c) they do not require frequent splitting and pruning and (d) the material could serve as an excellent form of nucleus stock to produce large number of plants rapidly (Bhojwani, 1981). Bajaj (1979) remarked that the highly sophisticated technique, involving "freeze-storage-thawing-culture, is a complicated multiple event and fault at any stage can result in lethal injury in the system". In contrast, the approach of storing cultures at low temperature (4–5°C) is very simple and has been used successfully for long term storage of many plants, e.g., grapes (12 months; Galzy, 1969); strawberry (6 years; Mullin and Schlegel, 1976) and apple (12 months; Lundergan and Janick, 1979).

Many of our orchids are endangered and require preservation through all possible means. In the present study, therefore, cold preservation of protocorms of *D. chrysanthum* and *P. spicerianum*, not tried earlier, was attempted.

MATERIALS AND METHODS

Protocorm development procedure:- Capsules of *D. chrysanthum* (330 days old) and *P. spicerianum* (270 days old) were collected from nature, surface sterilized by dipping in 90 per cent ethanol for a few seconds and flamed. The surface sterilized capsules were cut open with sterilized scalpel blades for extracting seeds. The extracted seeds of *D. chrysanthum* were sown *in vitro* on Knudson C semi-solid medium (Knudson, 1946). The seeds of *P. spicerianum* were sown on Thomale GD semi-solid medium (Thomale, 1954). The cultures of *D. chrysanthum* were incubated in continuous light (3,000 lux) at 24 ± 1°C. The cultures of *P. spicerianum* were first kept in dark and after 5 days of germination the cultures were kept in light for greening. Thirty-day-old green protocorms of *D. chrysanthum* and *P. spicerianum* were given low temperature treatment as described below:

Cold preservation procedure: Knudson C liquid medium and Thomale GD medium, both devoid of agar, having 2%, 4% and 6% sucrose were used for cold preservation of the protocorms. The pH of the medium was adjusted to 5.2. 4.5 ml of medium were poured into 5 ml borosil screwcap tubes. The tubes were autoclaved at 121°C and 15 lb/in² pressure for 20 minutes. Under aseptic conditions approximately 50 protocorms (length 0.35 mm, breadth 0.35 mm) were transferred into each tube containing liquid medium. For each set of sucrose concentration eighty tubes were maintained. Of these, 40 tubes were kept at 0°C and the remaining 40 tubes at 4°C.

Estimation of Plantlet Regeneration

For *D. chrysanthum:* Regeneration potential of the cold treated protocorms was compared using Knudson C (Kn C) and modified Knudson C (M Kn C; Bose and Bhattacharjee, 1980) semi-solid media having usual sucrose concentration (2%) and 5.2 pH.

For *P. spicerianum:* Regenerative potential of the cold treated protocorms was compared using Thomale GD (Th GD) and Thomale GD medium modified with addition of 0.3 mg 1^{-1} of niacin, thiamin HCl and pyridoxin HCl. 2% sucrose was incorporated in both Thomale GD and modified Th GD medium and pH adjusted to 5.0. The data on the change of colour of stored protocorms were recorded at the intervals of 3, 6 and 12 months. Five tubes from each treatment containing protocorms stored for 3, 6 and 12 months were cultured on different media and were kept under continuous light (3,000 lux) at 24 ± 1°C to investigate the growth responses of the regenerated plantlets. The characters taken into consideration were per cent plantlet regeneration after 180 days, the time taken for 1st leaf emergence and time taken for 1st root emergence.

Analysis of result: Students 't' test was used to test significance of plantlet regeneration percentage. Time taken in days was recorded for the 1st leaf and 1st root emergence.

RESULTS

Protocorms taken for the low temperature treatment were all green.

D. chrysanthum: Protocorms stored for 3 months at 0°C and 4°C in the Kn C liquid medium having 2% sucrose remained greenish. On the other hand, whereas the protocorms stored for 6 months at 4°C remained greenish, the protocorms stored at 0°C turned white. The protocorms stored for 12 months turned white irrespective of the storage temperature. When stored protocorms were cultured

Table 7.1 : Effect of cold preservation on protocorms of *D. chrysanthum*

Sucrose concentration	*Months stored*	*Temperature (°C)*	*Protocorm colour Kn C liquid medium*	*Plantlet regeneration (%) after 180 days*		*Time taken for 1st leaf emergence (days)*		*Time taken for 1st root emergence (days)*	
				Kn C	*M Kn C*	*Kn C*	*M Kn C*	*Kn C*	*M Kn C*
		0	Pale green	–	24.5 ± 1.74	–	40-70	–	–
2		4	Light green	-	53.0 ± 3.66	-	40-60	-	80-90
		0	Light green	-	46.5 ± 2.69	-	100-120	-	150-180
4	3	4	Light green	-	64.0 ± 3.05(a)	-	80-110	-	90-110
		0	Pale green	-	-	-	-	-	-
6		4	Light green	-	-	-	-	-	-
		0	White	-	-	-	-	-	-
2		4	Pale green	-	-	-	-	-	-
		0	Light green	-	39.0 ± 3.71	-	130-150	-	-
4	6	4	Light green	-	60.0 ± 2.35(a)	-	120-130	-	150-180
		0	Pale green	-	-	-	-	-	-
6		4	Pale green	-	-	-	-	-	-
		0	White	-	-	-	-	-	-
2		4	White	-	-	-	-	-	-
		0	Pale green	-	-	-	-	-	-
4	12	4	Pale green	-	-	-	-	-	-
		0	White	-	-	-	-	-	-
6		4	White	-	-	-	-	-	-

Value of plant regeneration are means (n = 10) ± S. E.

Significance tested by students 't' test (p = 0.05)

Treatments marked with same alphabet (a) not significantly different from each other.

on Kn C and MKn C medium, only protocorms stored for 3 months regenerated plantlets. Compared to protocorms stored at 0°C, the protocorms stored at 4°C revealed better regenerative potential and regenerated more plantlets (Table 7.1). Regeneration, however, occurred only in protocorms cultured on MKn C medium. The time required for the emergence of 1st leaf did not differ appreciably in plantlets regenerated from protocorms stored at 0°C or 4°C. The plantlets of protocorms stored at 4°C formed roots, but the plantlets regenerated from protocorms stored at 0°C did not develop any root.

Protocorms stored for 3, 6 and 12 months at 0°C and 4°C in the Kn C liquid medium having 4% sucrose remained greenish. When stored protocorms were cultured on Kn C and M Kn C medium, only protocorms stored for 3 months and 6 months regenerated plantlets while the protocorms stored for 12 months turned white. Compared to protocorms stored for 3 and 6 months at 0°C, the protocorms stored at 4°C regenerated more plantlets (Table 7.1). The time required for the emergence of 1st leaf and root differed appreciably in plantlets regenerated from protocorms stored for 3 months at 0°C and 4°C. However, in 6 months stored protocorms the time required for emergence of 1st leaf did not differ appreciably at the two storage temperatures (0 or 4°C). The plantlets of protocorms stored at 4°C for 6 months formed roots but the plantlets regenerated from protocorms stored at 0°C did not develop any root. Plantlets regeneration occurred only in MKn C medium. Protocorms of 12 months storage did not differentiate leaf and root.

Protocorms stored up to 6 months, at 0°C and 4°C in Kn C liquid medium having 6% sucrose remained greenish. The protocorms stored for 12 months turned white irrespective of storage temperature. When the protocorms stored for 3, 6 and 12 months in Kn C liquid medium having 6% sucrose at 0°C and 4°C were cultured on Kn C and MKn C semi-solid medium, no plantlet regeneration took place.

P. spicerianum: Protocorms preserved for varying time periods (3, 6 and 12 months) turned white. These protocorms when cultured on Th GD and modified Th GD medium did not regenerate.

DISCUSSION

The cold treated protocorms of *D. chrysanthum* regenerated plantlet only when cultured on MKn C semi-solid medium. Kn C medium was found unsuitable for regeneration of plantlet from cold treated protocorms. The MKn C medium differ from Kn C medium in the additional presence of vitamins (0.3 mg l^{-1} of niacin, thiamin HCl and pyridoxin HCl), macroelements (Mg SO_4 and KNO_3), microelements ($CoCl_2$ and KI) and in having lower concentrations of Ca $(NO_3)_2$, KH_2 PO_4 and NH_4 SO_4. The cold treated protocorms of *P. spicerianum*, however, lost their chlorophyll and did not become green when cultured on any of the medium. Lack of ability to develop chlorophyll may be the reason for loss of regenerative potential of stored protocorms of *P. spicerianum*.

With storage a change in colour, from green to white, of the protocorms of *D. chrysanthum* was observed during the low temperature treatments suggesting a decrease in chlorophyll content. Similar results have also been reported for *Cymbidium* protocorms cold preserved for 4–8 months at 1°C (Jacques *et al.*, 1982). Decrease in chlorophyll content in plants at lower temperature could be due not only to cold but also less light (Kuku/czanka *et al.*, 1990). 4% sucrose in Kn C liquid medium could cause maximum delay in the decrease of the protocorm colour. The loss of chlorophyll was less at 4°C in comparison to O°C. Compared to 2 and 6% sucrose; 4% sucrose was better in inducing regeneration of plantlets. Amongst the two temperatures (0 and 4°C) tested, 4°C was better for the regeneration of plantlets, emergence of 1st leaf and development of roots.

In the present study, *D. chrysanthum* protocorms preserved for a period upto 6 months at 4°C in Kn C liquid medium incorporated with 4% sucrose, stayed green and when cultured on M Kn C medium under normal cultural conditions, 60% protocorms regenerated plantlets. Thus, *D. chrysanthum* protocorms can be stored at 4°C for 6 months in Kn C liquid medium having 4% sucrose. Maximum regeneration of the protocorms can be achieved by further culturing them at 24 ± 1°C on M Kn C medium under continuous light as described above. However, attempt to preserve protocorms of *P. spicerianum* in the present study could not succeed.

8
Establishment of In Vitro Raised Seedlings in Field Conditions

Introduction

The plantlets produced *in vitro,* can not be transferred to *in vivo* environment straightaway due to resulting dessication (Corner and Thomas, 1981; Griffis et al., 1983; Dunstan and Turner, 1984). Therefore these are to be acclimatized before such transfer. Acclimatization is the process to make an organism to adapt to environmental changes (Brainerd and Fuchiagami, 1981). The acclimatization of micropropagated orchid seedlings from *in vitro* to *in vivo* environment is concerned with the transfer to : (a) humidity chamber; (b) to pots with compost. In the humidity chamber the seedlings transplanted in the community pots are sufficiently conditioned to withstand dessication. Subsequently community pots can be transferred to experimental net house/green houses.

In view of the above, an attempt was made to transfer and establish the *in vitro* raised seedlings of *D. chrysanthum* and *P. spicerianum* in the net house conditions.

MATERIALS AND METHODS

Capsules of optimum age (as in chapter 2) of *Dendrobium chrysanthum* and *Paphiopedilum spicerianum* were selected and germinated *in vitro* on basal medium as in chapter 2. The six months old seedlings from basal medium were subcultured on basal medium incorporated with 50 g l^{-1} banana homogenate (banana homogenate preparation was as in chapter 5). Six months later after the first subculturing, all uniform seedlings of *Dendrobium* and *Paphiopedilum* having 3-4 leaves and 3-5 roots were deflasked for hardening.

Hardening and acclimatization : The deflasked seedlings of *D. chrysanthum* and *P. spicerianum* were washed well with sterilized distilled water under aseptic conditions and transferred to 250 ml conical flask containing sterilized basal medium of 1/4 nutrient strength lacking in sucrose but having sterilized earthen pot pieces (1.0 cm size). Cotton plugs were used to close the mouth of the flasks. These flasks with transferred *in vitro* raised seedlings were maintained for 3 weeks on culture racks at $25\pm1^{\circ}$C temperature and 2,000-3,000 lux light.

After 3 weeks incubation the seedlings were taken out and transferred into various types of pot mixtures as detailed below :

(1) Brick pieces (approx. 5 mm × 5mm); charcoal pieces (approx. 5 mm × 5 mm); leaf mould; bark pieces (approx. 5 mm × 5 mm); tree fern (*Cyathea* spp.) pieces (approx. 5 mm × 5 mm); and dry *Sphagnum* in 1:1:1:1:1:2 ratio.

(2) Brick pieces (approx. 5 mm × 5 mm); charcoal pieces (approx. 5 mm × 5 mm); leaf mould; bark pieces (approx. 5 mm × 5mm); tree fern (*Cyathea* spp.) pieces (approx. 5 mm × 5 mm); dry *sphagnum* 1:1:1:1:1:2 ratio, with green *Sphagnum* topping.

(3) Leaf mould; perlite; vermiculite; dry *Sphagnum* in 1:1:1:2 ratio.

(4) Leaf mould; perlite; vermiculite; dry *Sphagnum* in 1:1:1:2 ratio with green *Sphagnum* topping.

(5) Seedlings were transferred to pots having plants of *Paphiopedilum*.

Plastic and earthen pots were taken and holes made underneath. The above pot mixtures (1-4) were then soaked in sterilized distilled water to make the compost wet. A tray with sterilized distilled water was kept under the pots. 3 pots each for the respective genus were taken. The potted seedlings were again maintained for 2 weeks on culture racks at 25 ± 1°C temperature and 2,000-3,000 lux light. After 2 weeks the potted seedlings were then transferred to humidity chamber (for a period of one month) and subsequently kept in net house conditions. Survival was estimated after 90 days from the day of transferring to humidity chamber.

Table 8.1: Survival of *D. chrysanthum and P. spicerianum* seedlings transferred to compost after 105 days (i.e., 2 weeks in culture racks + 1 month in humidity chamber + 2 months in net house).

		D. chrysanthum			*P. spicerianum*		
		Number of seedlings			*Number of Seedlings*		
	Composition of the compost	*Trans-ferred*	*Survived*	*Survival percentage (%)*	*Trans-ferred*	*Survived*	*Survival percentage (%)*
1.	Brick pieces, charcoal pieces, bark pieces, leaf mould, tree fern pieces, dry *Sphagnum* in 1:1:1:1:1:2 ratio	15	15	100	7	0	0
2.	Above compost with green *Sphagnum* on top	15	15	100	7	0	0
3.	Leaf mould, perlite, vermiculite, *Sphagnum* (dry) in 1:1:1:2 ratio	15	0	0	7	6	85.7
4.	Above compost with green *Sphagnum* on top	15	0	0	7	4	57.1
5.	Nearadult plants	15	0	0	7	7	100

Values are means (n=3)

RESULTS

The seeds of *Dendrobium chrysanthum* germinated and differentiated seedlings on Knudson C basal medium while seeds of *Paphiopedilum spicerianum* germinated and differentiated seedlings on Thomale GD basal medium. The six month old *in vitro* raised seedlings of *Dendrobium* and *Paphiopedilum* when subcultured on respective basal medium having banana homogenate developed further which could be transferred to community pots in the next six months *(Plate 8.1* and *8.2)*.

In vitro raised seedlings transferred directly to net house conditions in the various composts without undergoing hardening and acclimatization did not survive. However, seedlings which underwent hardening and acclimatization revealed survival in some composts *(Plate 8.3-8.8)*. Table 8.1 gives data on effect different compost on survival of the seedlings of the epiphytic species *D. chrysanthum* and terrestrial species *P. spicerianum*. *D. chrysanthum* exhibit 100% survival on potting medium containing brick pieces, charcoal pieces, bark pieces leaf mould, tree fern (*Cyathea* spp.) pieces and dry *Sphagnum* in the ratio of 1:1:1:1:1:2; with or without green *Sphagnum* on the surface of the compost (Table 8.1; *Plate 8.3-8.5*). No seedling survival was evident on the other three composts (Table 8.1). *P. spicerianum* seedlings did not survive on composts suitable for *D. chrysanthum* seedlings. *P. spicerianum* seedlings revealed 85 per cent survival on leaf mould, perlite, vermiculite and dry *Sphagnum* in 1:1:1:2 ratio; 57 per cent survival was recorded when green *Sphagnum* was placed on the top of the above compost (Table 8.1; *Plate 8.6 and 8.7)*. However, 100 per cent survival of *P. spicerianum* seedlings was observed when *in vitro* raised seedlings were transferred to pots having plants of *Paphiopedilum* (Table 8.1, *Plate 8.8*).

DISCUSSION

In the present study it was observed that for establishment of *in vitro* raised seedlings of *D. chrysanthum* and *P. spicerianum* in field conditions acclimatization of the seedlings was essential because otherwise the seedlings do not survive. Acclimatization of the *in vitro* raised seedlings is important to minimize water stress (Mohammed and Vidaver, 1988). However, studies on *Phalaenopsis* hybrid (Seeni and Latha, 1990) and *Dendrobium* hybrid (Kumar, 1992) reveal that deflasked seedlings do not require acclimatization. The seedlings easily got established in the community pots within 2 weeks after transplantation (Seeni and Latha, 1990). In the present investigation, however, it was not so because the *in vitro* raised seedlings of *D. chrysanthum* and *P. spicerianum* had to be acclimatized prior to transferring to field conditions. The performance of the seedlings of the epiphytic species *D. chrysanthum* and terrestrial species *P. spicerianum* varied with the nature of the potting medium. Of the various potting media tried brick pieces, charcoal pieces, tree fern (*Cyathea* spp.) pieces, bark pieces, leaf mould and dry *Sphagnum* in 1:1:1:1:1:2 ratio with and without green *Sphagnum* on top supported the maximum survival for the epiphytic species *D. chrysanthum*. *D. chrysanthum* seedlings did not survive on the potting medium suitable for *P. spicerianum*. In case of *P. spicerianum* maximum survival was observed with

Plate 8 : Establishment of *in vitro* raised seedlings in community pots

1. *in vitro* raised seedlings of *D. chrysanthum* ready for transfer to community pots
2. *in vitro* raised seedlings of *P. spicerianum* ready for transfer to community pots
3. seedlings of *Dendrobium* growing in compost of brick pieces, charcoal pieces, bark pieces, leaf mould, tree fern and dry *Sphagnum*, 105 days old
4. seedlings of *Dendrobium* growing on compost of brick pieces, charcoal pieces, bark pieces, leaf mould, tree fern, dry *Sphagnum* with green *Sphagnum* on top, 105 days old
5. *Dendrobium* seedlings two years old in community pots
6. seedlings of *Paphiopedilum* growing in compost of leaf mould, perlite, vermiculite, dry *Sphagnum*, 105 days old
7. seedlings of *Paphiopedilum* growing in compost of leaf mould, perlite, vermiculite, dry *Sphagnum* with green *Sphagnum* on top, 105 days old
8. seedlings of *Paphiopedilum* growing near adult plants, 105 days old

1
2
3
4
5
6
7
8

seedlings transferred to pots containing established adult *Paphiopedilum* plants followed by compost containing leaf mould, perlite, vermiculite and dry *Sphagnum* in 1:1:1:2 ratio; leaf mould, perlite vermiculite, dry *Sphagnum* in 1:1:1:2 ratio with green *Sphagnum* on top. The seedlings of *Paphiopedilum* did not survive on potting medium suitable for *D. chrysanthum.*

Thus, in the present investigation acclimatization of the *in vitro* raised seedlings of both the genus was essential for transferring them to community pots. The present investigation thus supports the conclusions of Corner and Thomas (1981), Dunstan and Turner (1984) and Griffis *et al.*, (1983).

9
Commercial Cultivation of Orchids

Introduction

Orchids, because of their ornamental value, are much in demand in both domestic and international markets. India is far behind Singapore, Thailand, Malaysia, Australia etc. in commercial cultivation of orchids mainly because commercial cultivation of orchid is still in infancy in India. The major developments in orchid cultivation, in the world, have been due to modern scientific technology including embryo/seed and meristem culture.

Majority of orchids thrive well in humid tropical and subtropical climate with medium intensity of sunlight. Due to ever increasing demand of orchids in the International market there is great potential for orchid industry in India. Arunachal (Orchid Research Centre), Assam (India Carbon, Guwahati), Meghalaya (in collaboration with Singapore firm at Barapani), Sikkim, West Bengal (Kalimpong and Darjeeling) and Kerala (Trivandrum to Kannur; AVT-Vanitha Club) are excellent examples of this. Considering the importance, potential and feasibility of orchid Industry in India the Govt. of India and its agencies have come forward with schemes of financial assistance for orchid cultivation.

Cultivation

Various aspects of mass multiplication of epiphytic *(Dendrobium)* and terrestrial *(Paphiopedilum)* orchids, using tissue culture techniques, and their establishment have been dealt in the earlier chapters.

Orchids can be cultivated outdoor as well as indoors. Outdoor cultivation of orchids can be achieved only at places having ideal agroclimatic conditions. Epiphytic orchids can be grown well on old well grown trees of oak, mango, jamun, *Bauhinia* (Kachnar group) and other rough barked trees, which are good host trees for orchids in N.E. India and eastern and western ghats of India. Terrestrial orchids, on the other hand, can be grown on grounds by making nursery beds.

Indoor Cultivation

(a) *Agro-shade net houses :* Orchids can be cultivated indoor in orchid houses (Fig. 9.1) constructed with wood/bamboo. The roof of these orchid houses is covered with agro-shade nets (black with 50 per cent shade) while the sides are kept open or covered with wire net mesh.

(b) *Green house/glass house :* The success of growing healthy orchids is dependent on availability of proper conditions of growth (humidity, light, temperature, ventillation, water and nutrients). The optimal conditions of growth could

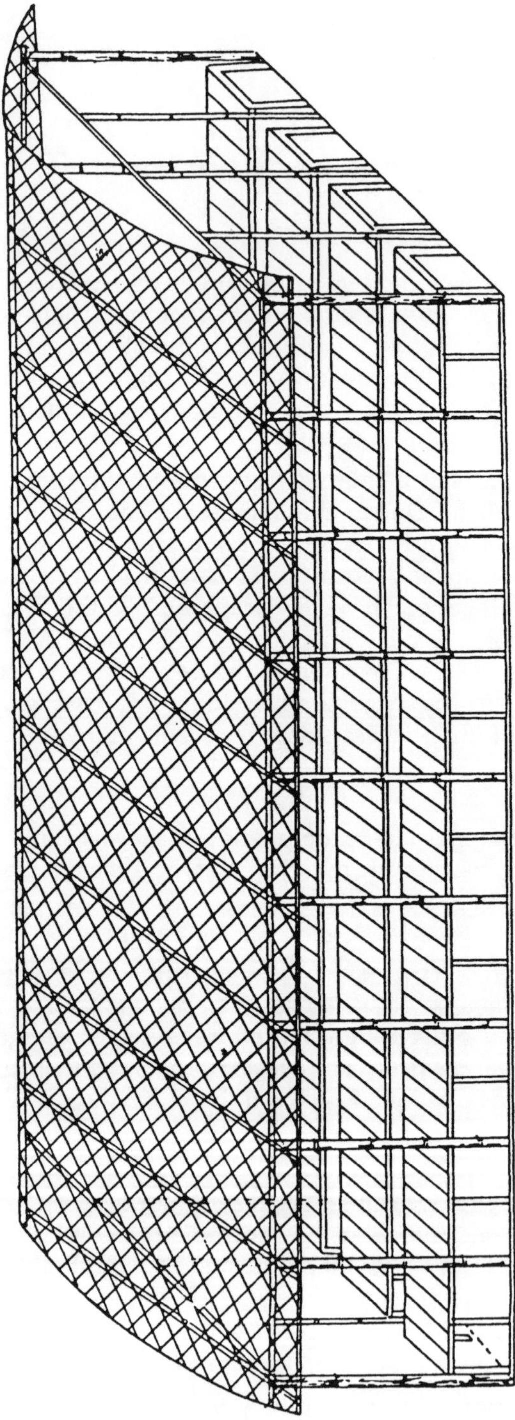

Fig. 9.1 : An agro-shade net house with 4 stages

easily be created in a green house (i.e., the glass house). Green houses are essentially for growing orchids in the plains of India. However, maintenance of the adequate conditions for orchid cultivation especially humidity and temperature, in the plains would be expensive.

The glass house should be south facing and oriented east-west. Some deciduous trees can be grown on the southern side for giving shade during late spring (when many tender young outgrowths are present) and summer time. Shade can also be provided by fitting movable blinds. An airspace between the glass and blinds is essential to maintain more uniform temperature in the house by allowing free circulation of air over the glass. From early autumn more light is to be admitted to ripen the bulbs making the plant firm. *Dendrobium* require abundant light to make them flower while *Paphiopedilum* are shade loving.

The size of the glass house should not be less than 100 sq. ft. i.e., approx. 8 x 12 ft. with staging on two sides and a $2\frac{1}{2}$ ft. path in between. The height of the glass house should be 8 ft. For providing a centre staging to incorporate taller plants the minimum size of the glass house should be 16 x 24 ft. (Fig. 9.2). Three tired staging can be used on either sides (Fig. 9.2), this would enable a great number of plants to be accommodated. Whereas more sunloving plants should be put on the upper shelf, the less light loving plants can be kept on the lower shelves.

Humidity can be maintained either by keeping water fillfed trays below the lower most shelf or by placing water retaining material on the floor itself below the shelves; which when kept moist will maintain humidity. Some of the automatic humidifier (costing approx. Rs. 8000) can also be used. The humidity in glass house, however, should not be less than 30 per cent in night and above 80 per cent during day time.

Cultivated orchids require plenty of fresh air. Thus for providing an ample flow of air around the plant and for regulation of temperature, ventilrrators should be provided near floor and roof of the green house (Fig. 9.2). However, draughts should be avoided. Exhaust fans can be fixed for more flow of air and cooling of glass house. In orchids there are three section, the *warm section* (with summer night temp. of about 21°C and day temp. 21-27°C; higher during the brighter spells; and in winter 18°C by night and 21°C by day), the *intermediate section* (with summer night temp. of about 18°C and day temp. 18-21°C; higher during the brighter spells; and in winters 16°C by night and 16-18°C by day) and the *cool section* (with summer night and day temp. of about 16°C; and in winters down to about 10°C and 13-16°C by day).

Watering of the orchids is important. During the main growing season orchids shall be watered 2-3 times a week, but in winters watering should either be done once a week or it could be further delayed depending on the species since high humidity around the plant reduces the rate of water loss through transpiration.

Nutrient feeding is essential for orchids in green house. Atleast once a week nutrient feeding should be given.

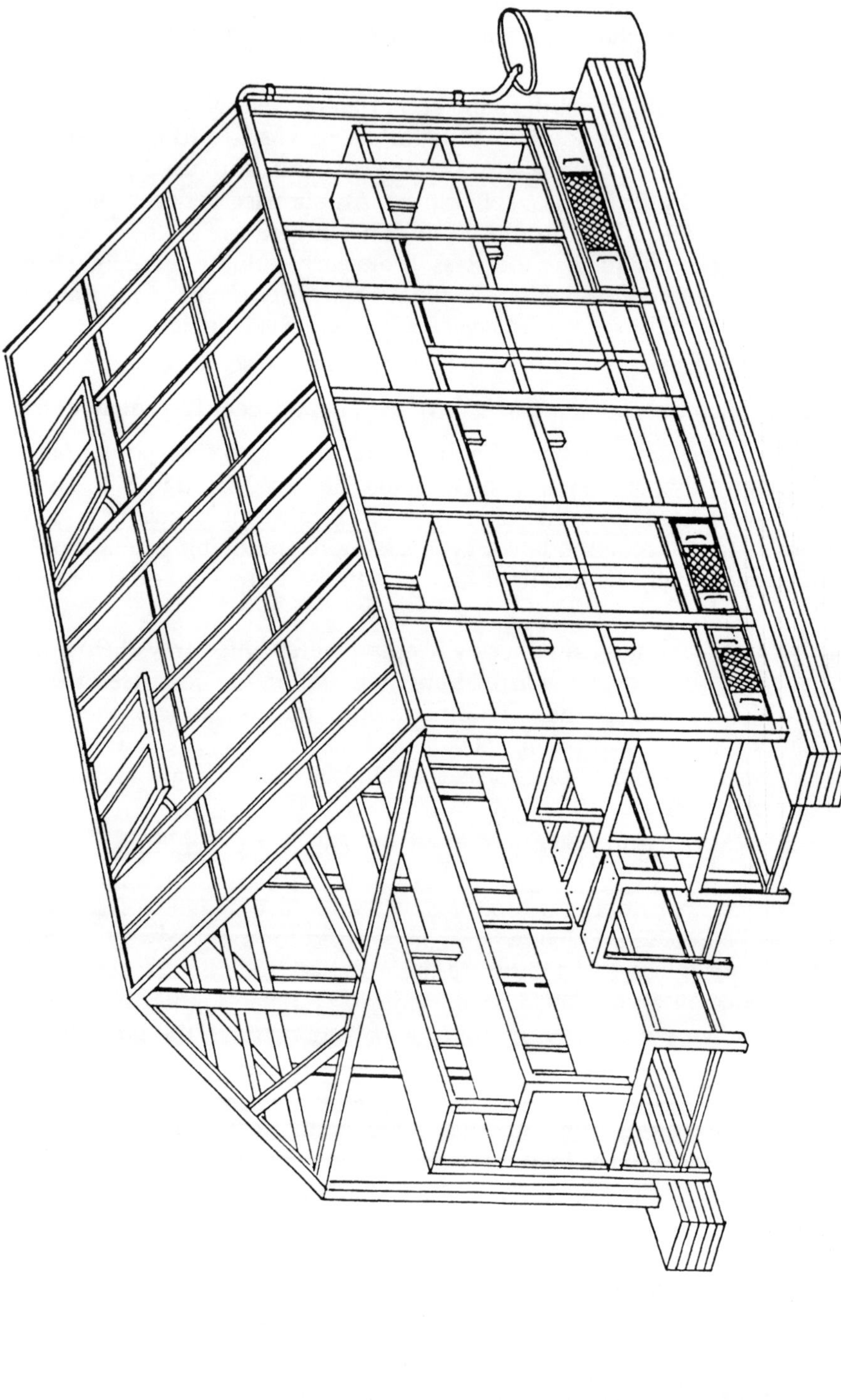

Fig. 9.2 : A green house with three staging, two on either sides and one on the centre; ventilation on the roof top and at the base.

Propagation

Depending on the growth habit i.e., sympodial or monopodial the cultivated orchids are propagated by splitting or top cutting respectively. The sympodial orchids e.g., *Paphiopedilum, Dendrobium, Cymbidium* etc., while being splitted should have atleast two plants/pseudobulbs. The cut faces of the split plant are dusted with a fungicide to prevent fungal infection and rotting. All dead roots are being trimmed and are then potted. In monopodial orchids e.g., *Vanda, Ascocentrum, Phalaenopsis* etc., which are propagated by top cutting, the stump also sends up one or more shoots. Some orchids reproduce by means of keikis i.e., plantlets that grow from the nodes of pseudobulbs, sometimes from flower stalks and rarely from the roots. All Dendrobiums are capable of reproduction from nodes of pseudobulbs.

Terrestrial orchids like *Paphiopedilum, Cymbidium* etc., are potted in a porous compost. The compost is a mixture of coarse sand, leaf mould, soil, dry cowdung powder (optional) rice husk/dry crushed sphagnum (optional) in 2:1:1:1:1 ratio. The pots used for potting should be holed at the base and crocked to half of its depth. The plant is then placed and potted with the compost to a depth sufficient to cover the base or the crown of the roots.

The epiphytic orchids are aerial feeders. Therefore, for potting epiphytic orchids adequate ventillation and drainage is provided (Fig. 9.3). The pots are holed (1 cm diam). on the sides in addition to the bottom drainage hole. The pots are crocked to half of its depth. The plant is then placed in the pot and covered upto the base of the root with a compost of pine bark, charcoal and rock pieces in 3:1:1 ratio; one part of tree fern can also be used. Size of the chunks of the potting medium is according to the size of the pots used for growing the required plant.

Repotting is undertaken during the spring months of the year, when the plant is commencing new growth.

Pests and Diseases

A clean, airy environment with right humidity can prevent plants from being attacked by pests. A constant and attentive watch can control and eradicate them.

Mealy bugs and scale insects : The sap sucking pests attack Dendrobiums having soft canes e.g., *D. nobile*. Since the insects hide in the axils of the leaves of orchids, the dry bracts, leaves and flower, stem, no longer in bloom, should be removed. Systemic insecticide which makes plant toxic to sap sucking pests should be applied 2-3 times at 10-14 days interval for controlling the pests.

Red spider mite : These also suck the sap by feeding on the underside of the leaf. Usually soft leaved genera e.g., *Calanthe, Cymbidium* are susceptible. The infected leaves exhibit slight mottling on the underside. But in severe infection leaves turn yellow and reveal a fine gossamer web on the underside. Humid conditions prevent the attack. Use of Malathion, Kelthane, Dimite etc., is an effective control. For the active colonies 3-4 applications at 8-10 days interval should control.

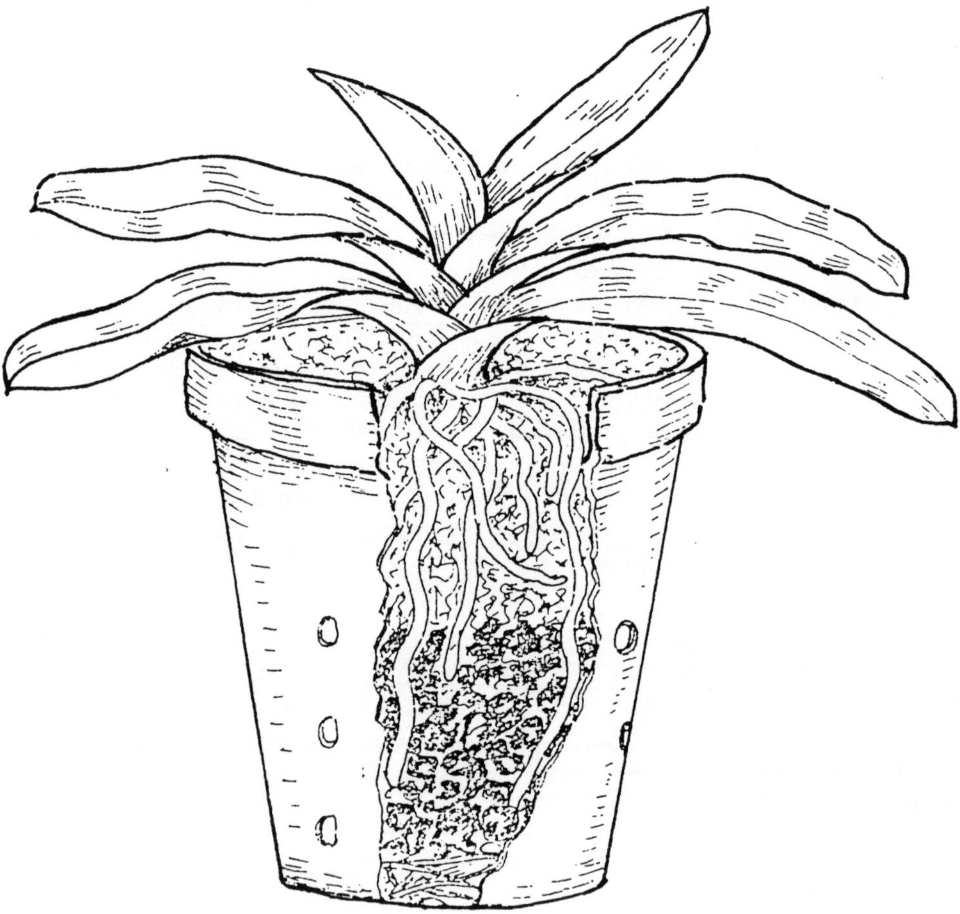

Fig. 9.3 : A pot section showing potting medium

Aphids: They commonly attack the flower buds, young growth and stems, and distort flowers. Here again the use of malathion, kethane, dimite etc. should be effective.

Thrips : They puncture the upper surface of the leaves and cause silvery areas. They can be controlled by parathion, benzene or DDT spray.

Slugs and snails : In a single night they can destroy inflorescence, a young leaf or newly forming root. Pseudobulb rupture are also caused by small snails and slugs that eat into and eventually hollow out the bulb. Sprinkling of metaldehyde periodically over floor and benches would help clearing slugs and snails. They should also be handipicked and destroyed.

Leaf hoppers and beetles : They sometimes damage green leaves and flowers; and can be controlled by malathion spray.

Cockroaches : They eat up petals and flower spikes. Destroying their breeding places is the best eradication measure.

Millipeds : They can be controlled by dieldrin spray.

Fungal and Bacterial Diseases

The main sources for these diseases are benches, pots, potting composts and the implements. All these things should be properly sterilized using disinfectants such as formalin or lime sulphur. Standing water, improper ventillation, excessive humidity help in spreading this disease. Brown patches of soft tissue, which grow bigger very quickly on leaves or pseudobulbs are indications of fungal or bacterial disease.

Fungal diseases like sooty mould that cover the aerial parts of stem can be removed by cotton dipped in methylated spirit or fungicide; rust fungus forming the spots which turn reddish brown or finally black can be eradicated with the use of copper fungicide, tersan or mercuric chloride; the heart rot disease which develops as dark, sharply delimited lesions on the plant, spreading rapidly causing yellowing of leaves and finally leaf fall, treatment with tersan dust checks this disease.

Bacterial diseases are like the soft rot disease causing slimy rot of pseudobulbs and stem bases, it spreads rapidly, destroying the effected plants, spraying the surrounding plants with 10 ppm mercuric chloride can save the plants. Bacterial brown spots is another severe disease which can also be controlled by mercuric chloride treatment.

Virus Disease

Any tiny spot on leaves, leaf necrosis, any alteration in colour of the flower tends to be diagonised as virus attack. Virus disease are incurable. So, if symptoms of virus are detected the plants should be removed immediately and destroyed.

Table 9.1: Unit cost for cultivating 1000 plants of ornamentally important indigenous terrestrial orchid species **Paphiopedilumi* epiphytic species of *Dendrobium* in pots under **Agro-shade net house by small farmers.

	I Yr.	*II Yr.*	*III Yr.*	*IV Yr.*	*V Yr.*	*VI Yr*	*Total*
I. **INVESTMENT**							
A. **Agro-shade net house** (24' × 36') with ± 1368 pot capacity							
Wood/bamboo frame	5,000	—	500	500	500	500	7,000
Agro-shade net for covering top	1,700	—	—	—	—	—	1,700
Single staging : 4 numbers	4,000	—	—	—	—	—	4,000
B. **Potting material**							
1000 pots at Rs. 3 each & from 3rd year, purchase of 500 pots yearly	3,000	—	1,500	1,500	1,500	1,500	9,000
Plant material at ± Rs. 20 each	20,000	—	—	—	—	—	20,000
Potting mixture	5,000	—	1,000	1,000	1,000	1,000	9,000
Manure & fertilizer	1,000	—	500	500	500	500	3,000
C. **Plant protection and manual irrigation**							
Equipments	4,500	—	—	—	—	—	4,500
Chemicals	300	—	300	300	300	300	1,500
D. **Post harvest handling & packaging**	—	—	1,500	1,500	1,500	1,500	6,000
E. **Labour charge**	7,200	7,200	12,000	12,000	12,000	12,000	62,400
F. **Total investment**	51,700	7,200	17,300	17,300	17,300	17,300	1,28,100
II. **SALES REALISATION** (Lowest side)							
A. Cut flowers (no. of spikes)	—	—	2,000	2,000	2,000	2,000	
Cost per spike ± Rs. 5	—	—	10,000	10,000	10,000	10,000	40,000
B. Number of vegetatively multiplied plants for sale	—	—	2,000	2,000	2,000	2,000	
Cost per plant ± Rs. 25	—	—	50,000	50,000	50,000	50,000	2,00,000
C. Total Return	—	—	60,000	60,000	60,000	60,000	2,40,000
III. MINIMUM NET PROFIT (II C - I F)	—	—	42,700	42,700	42,700	42,700	1,11,900

**P. fairrieanum, P. hirsutissimum, P. spicerianum, P. venustum, P. druryii*; For Tropical & Sub-tropical agroclimatic zone.
P. insigne, P. villosum; For temperate and cold Sub-tropical zone.
** 50% shade agro-net of black colour is recommended.
Thus, from the 6th year there is net recurring profit of Rs. 42,700 per annum from the growing of 1000 plants.

Table 9.2: Unit cost for setting up an orchid industry

	I Yr.	II Yr.	III Yr.	IV Yr.	V Yr.	VI Yr.	Total
I. **INVESTMENT**							
A. LABORATORY							
Construction							
20' × 48' (2 rooms: i. experimental laboratory ii culture room)	2,88,000	—	—	—	—	—	2,88,000
Equipments (Laminar flow B.O.D., double distillation unit, balance, pH meter, culture racks, air-conditioner, humidifier etc.)	2,85,000	—	—	—	—	—	2,85,000
Chemicals	25,000	2,000	2,000	2,000	2,000	2,000	35,000
Glasswares	60,000	20,000	20,000	20,000	20,000	20,000	1,60,000
Glass bottles (5000 numbers) (wide mouth bottles, e.g. saline bottles, horlicks bottles etc. for subculturing the *in vitro* rasied seedlings from IInd year at the rate of Rs. 1 per bottle)	—	5,000	50,000	5,000	5,000	5,000	25,000
Skilled man power (1 number) Rs. 3000 per month	36,000	36,000	36,000	36,000	36,000	36,000	2,16,000
Labour charge (1 number) Rs. 1000 per month	12,000	12,000	12,000	12,000	12,000	12,000	72,000
B. GREEN HOUSE							
2 numbers, 16' × 24' each (complete with three staging on either side and single staging centre, sprinkler, humidifier, chiks etc.) capacity approx, 1368 pots	5,00,000	—	—	—	—	—	5,00,000

(Contd.)

		I Yr.	II Yr.	III Yr.	IV Yr.	V Yr.	VI Yr.	Total
	C. AGRO-SHADE NET HOUSE							
	24' × 36' (with wooden/bamboo frame, agro-shade nets, benches etc.) capacity ± 1368 pots	11,000	1,500	1,500	1,500	1,500	1,500	18,500
	D. POTTING MATERIAL (for both B and C)							
	Cost for 2000 pots at Rs. 3 each & from 2nd Yr., purchase of 300 pots yearly.	6,000	900	900	900	900	900	10,500
	Cost of plant material ± Rs. 20 each	40,000	—	—	—	—	—	40,000
	Cost for 700 pots at Rs. 3 each to grow the laboratory raised seedlings.	—	—	—	2,100 2,100	2,100	2,100	6,300
	Potting mixture	6,000	—	2,000	2,000	2,000	2,000	14,000
	Manure and fertilizer	1,500	—	1,000	1,000	1,000	1,000	5,500
	E. PLANT PROTECTION AND MANUAL IRRIGATION							
	Equipment	4,500	—	—	—	—	—	4,500
	Chemical	300	—	300	300	300	300	1,500
	F. POST HARVEST HANDLING AND PACKAGING	—	—	6,000	6,000	6,000	6,000	24,000
	G. LABOUR CHARGE (2 numbers)	24,000	24,000	24,000	24,000	24,000	24,000	1,44,000
	H. ELECTRICITY	21,600	21,600	21,600	21,600	21,600	21,600	1,29,600
	I. TOTAL INVESTMENT (Rs.)	13,20,900	1,23,000	1,32,300	1,34,400	1,34,400	1,34,400	19,79,400
II.	**SALES REALISATION**							
	A. LABORATORY							
	Sale of 250 flasked seedlings per month at the rate of ± Rs. 150 per flask.	—	—	4,50,000	4,50,000	4,50,000	4,50,000	18,00,000

(Contd.)

	I Yr.	II Yr.	III Yr.	IV Yr.	V Yr.	VI Yr.	Total
B. GREEN HOUSE AND AGRO-SHADE							
NET HOUSE							
Established laboratory raised seedlings							
(a) number of plants for sale	—	—	—	—	10,500	10,500	
Cost per plant ± Rs. 15 in Vth year & Rs. 25 from 6th year onwards	—	—	—	—	1,57,500	2,62,500	4,20,000
(b) Cost of cut flowers from 10,500 established seedlings	—	—	—	—	—	52,500	52,500
From plants purchased							
(c) number of vegetatively multiplied plants for sale	—	—	4,000	4,000	4,000	4,000	
cost per plant ± Rs. 25	—	—	1,00,000	1,00,000	1,00,000	1,00,000	4,00,000
(d) cut flowers (number of spikes)	—	—	4,000	4,000	4,000	4,000	
cost per spike ± Rs. 5	—	—	20,000	20,000	20,000	20,000	80,000
C. TOTAL RETURN	—	—	5,70,000	5,70,000	7,27,500	8,85,000	27,52,500
III. **NET PROFIT (IIc - I)**	—	—	4,37,700	4,35,600	5,93,100	7,50,600	7,73,100

Thus, from this establishment there is a net recurring profit of Rs. 7,50,600 / per annum from the 6th year.

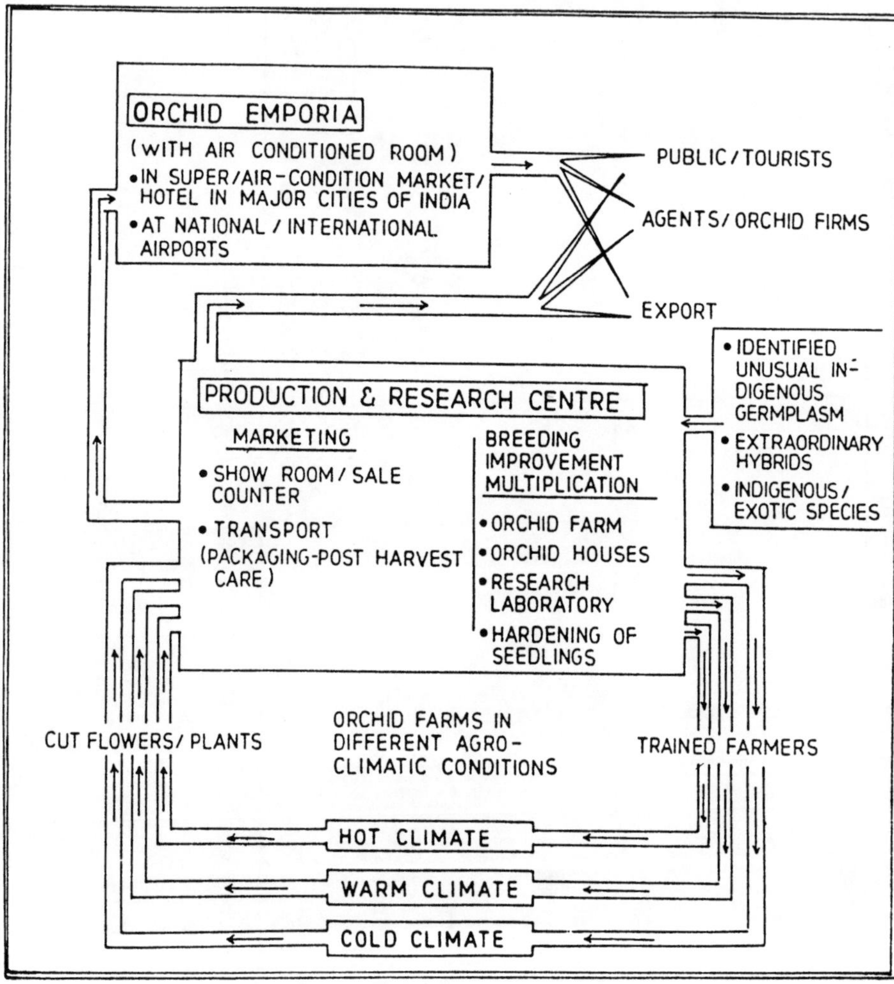

Fig. 9.4 : Steps for a viable orchid industry in India

Orchid cultivation as cottage industry/an industry

Unit cost for cultivating 1000 plants of ornamentally important indigenous terrestrial orchid species, *Paphiopedilum*/epiphytic species of *Dendrobium* in pots under Agro-shade net house by small farmers is given in Table 9.1. Thus from the 6th year there is net recurring profit of Rs. 42,700 per annum. The estimated returns from the setting up of an orchid industry which would include laboratory, green house and agro-shade net house are presented in the Table 9.2. From the 6th year there is a net recurring profit of Rs. 7,50,600 per annum. Steps for setting up of a viable orchid industry in India is given in Fig. 9.4.

10
General Discussion and Conclusions

Orchid seeds are produced in large numbers within a capsule. The seeds are very minute and lack metabolic machinery and functional endosperm with the result that only 0.2-0.3 per cent of the seed germinate in nature (Singh, 1992). In 1922, Knudson demonstrated that it is possible to germinate orchid seeds under aseptic conditions using sugar enriched mineral medium.

In the present investigation the stage of maturation of seed influenced the rate of *in vitro* germination and the stage at which seeds could be cultured successfully varied with the genus and the species used. Seeds of Dendrobiums germinated earlier than Paphiopedilums. Within the genus *Dendrobium* germination occurred first in *D. aphyllum, D. chrysanthum, D. falconeri* and *D. wardianum* and last in *D. ochreatum*. The seeds of *D. chrysotoxum, D. heterocarpum* and *D. primulianum* were intermediate in germination. Within the genus *Paphiopedilum*, the seeds of *P. insigne, P. spicerianum* were first to germinate and seeds of *P. hirsutissimum, P. venustum* and *P. villosum* germinated last. The seeds of *P. fairrieanum* were intermediate in germination. Arditti *et al*., (1982) suggested that the earliest stage at which embryos can be cultured successfully varies with genus, species, hybrids and local conditions. In the present investigation the time required for germination by the different species of both the genus is species dependent.

In all the species of *Dendrobium* and *Paphiopedilum* seeds from 90 days old capsules never germinated. In *Dendrobium*, except *D. aphyllum* and *D. falconeri* and in *Paphiopedilum* except for *P. spicerianum* seeds from 150 days old capsules did not germinate. In seeds obtained from 210 days old capsules in most of the species germinated (*D. chrysanthum, D. chrysotoxum, D. herterocarpum* and *D. wardianum*) except *D. ochreatum* and *D. primulianum* where seed germination occurred only after the capsules were 270 days old. In *Paphiopedilum* seed obtained from 210 days old capsules in *P. insigne, P. venustum* and *P. villosum* germinated and in *P. fairrieanum, P. hirsutissimum* seeds germination occurred only after the capsule were 270 days old. These findings suggest that developing embryo in different species reaches the critical stage of development necessary for seed germination at different times. Similarly, there are reports on other species also for the critical stage of development necessary for seed germination viz., 120 days for *Dendrobium nobile* (Ito, 1955); 50 days for *Vanda* Miss Joaquim (Rao and Avadhani, 1964; Ong, 1969); 60 days for *Vanilla* (Withner, 1955) etc. In both Dendrobiums and Paphiopedilums usually seed germination increased with increasing age of the capsule upto a threshold and then decreased. Assimilation of inhibitors in seed coat (Burgeff, 1954) in addition to onset of dormancy (Vij *et al*., 1981) could be the cause for the decrease in germination of seeds from older capsules.

Thus the capsule age is very important factor for achieving optimum seed germination and it is species specific.

Seeds of both the taxa germinated and devėloped seedlings on various media used. However, in *D. chrysanthum* (epiphytic species) Knudson C medium having inorganic ion concentration 29.72 millimoles and in *P. spicerianum* (terrestrial species) Thomale GD medium having inorganic ion concentration 24.3 millimoles proved better for the germination of seeds and seedling growth. Arditti and Ernst (1984) pointed out that orchid seed germinates on a wide variety of inorganic salt combinations and ion concentrations, but in general a balanced mineral mixture and medium concentration is an important factor for every orchid species. Thus, the epiphytic species *D. chrysanthum* prefers a medium having higher ion concentration compared to the terrestrial species *P. spicerianum*. Arditti and Ernst (1984) have pointed out that for terrestrial orchid, especially lady's slipper (*Cypripedium, Paphiopedilum*), medium with low ion concentration i.e., a more dilute medium is better.

The epiphytic species *D. chrysanthum* germinated both in light and dark. But in case of the terrestrial species *P. spicerianum* light proved inhibitory for germination of seeds, the seeds germinated only in dark. Arditti (1967, 1979) and Arditti *et al.*, (1981) also pointed out that epiphytic species germinated both in light and dark; and Kano (1965), Harvais (1972) and Stoutamire (1974) pointed out light to be an inhibitory factor in some terrestrial species. However, in both the species darkness had an inhibitory effect on pigmentation, differentiation and survival of seedlings. Therefore, eventhough the seeds of *P. spicerianum* germinated only in dark, just after seed germination, the germinated seeds needs to be transferred to light for proper development and survival of seedling. Thus, in the present investigation while the epiphytic species is not specific in its light requirement for germination, the terrestrial species is very specific in its light requirement since their seeds germinated only in dark. Light in both the taxa was essential for chorophyll, shoot and root development and for survival of the seedling in both the taxa.

In the present investigation, moderately acidic medium (pH 5) was most suitable for seed germination and seedling growth in both the taxa. Suitability of medium having pH 5.0 could be related to regulating the availability of compounds, as lower pH causes increased availability of minor elements such as iron and maganese and higher pH causes the precipitation of important inorganic nutrients (Vacin and Went, 1949 and Withner, 1959). pH 4.8 to 5.2 being optimum pH for orchild seed germination has also been reported by Arditti (1967, 1979) and Arditti and Ernst (1984).

Incorporation of carbohydrate sources in both the genera was essential for seed germination. *D. chrysanthum* seeds exhibit very poor germination on medium devoid of carbohydrate sources and differentiation beyond the protocorm stage did not occur. In *P. spicerianum* seeds did not germinate at all on medium lacking in carbohydrate. This requirement of carbohydrate could be due to lack of endosperm, the tissue which nourishes developing embryo, in the orchid seeds. In both *D. chrysanthum* and *P. spicerianum*, all the carbohydrates with the exception of lactose and soluble starch proved suitable as sources of carbohydrate for seed germination and subsequent growth. However, sucrose 2 per cent was most suitable for inducing enhanced seed germination and better seedling differentiation in both the genera.

For seed germination and seedling growth incorporation of nitrogen source in the medium was also essential. In the absence of any nitrogen source *D. chrysanthum* seeds exhibited poor germination and germinated seeds did not differentiate beyond the protocorm stage. No germination of *P. spicerianum,* seeds, however, occurred in the absence of nitrogen. Unlike organic nitrogen (L-arginine, L-aspartic acid, L-phenylalanine, glycine) inorganic nitrogen, when incorporated in the basal medium, induced seed germination and growth. Thus for inducing germination of seeds and growth, compared to ammonium salts, nitrate salts were more effecitive in *D. chrysanthum*. However for *P. spicerianum* ammonium salts were effective specifically. Burgeff (1936) also stated that nitrate as nitrogenous source was best used by epiphytic orchid species, but those species which germinate in dark require ammonium salts specifically as nitrogenous source. In the present investigation Ca $(NO_3)_2$ caused maximum seed germination and development of well differentiated chlorophyllous seedling in *D. chrysanthum* while NH_4 SO_4 was most suitable for seed germination and subsequent seedling growth in *P. spicerianum.*

Amongst the growth hormones, kinetin eventhough did not improve germination of *D. chrysanthum* seeds it induced better shoot and root growth. BAP in the medium, on the other hand, caused unsatisfactory germination and growth. Auxins also inhibited seed germination and growth in *D. chrysanthum*. These findings are in agreement with Kano (1965) and Miyazaki and Nagamatsu (1965). In *P. spicerianum* cytokinin had no beneficial effect. However, auxin supplemented medium with the exception of 2, 4-D revealed a speedy and increased seed germination compared to control. The results support the findings of Arditti and Ernst (1984). However, further growth of the germinated seeds of *P. spicerianum* was inhibited in auxin supplemented medium. GA_3 had an inhibitory effect on seed germination in both the genera. With the incorporation of GA_3 in the medium subsequent growth remained unaffected in *D. chrysanthum* while it was inhibited in *P. spicerianum*. Thus the study reveals that auxin, with the exception of 2, 4,-D induced speedy and increased seed germination in *P. spicerianum* but inhibited seed germination in *D. chrysanthum*. However, in both the genera subsequent growth of the germinated seeds on the auxin supplemented medium was inhibited. Kinetin induced better shoot and root growth in *D. chrysanthum* but in *P. spicerianum* kinetin had no beneficial effect.

Organic supplements inhibited germination of seeds in both the genera. However, seedling differentiation and development of pigments remained unaffected with CM and activated charcoal. The present investigation, suggest that both *D. chrysanthum* and *P. spicerianum* do not require the incorporation organic supplement for seed germination and subsequent differentiation.

In the investigations with clonal propagation Kn C medium proved better for the explants of *D. chrysanthum*. However, differentiation of shoots from various explants was better when Kn C medium was modified. On Kn C medium, supplemented with CM, shoot tip explants developed protocorm-like bodies which later produced shoots. Wilfret (1966) pointed out that in *Cymbidium* hybrids internodal areas between the embryonic leaves frequently expand and differentiate protocorm-like bodies. Addition of activated charcoal to the Kn C medium supplemented with CM induced callus formation on root tip explants of *in vitro* seedlings which later developed shoots. Root tip culture has been reported to be very useful as multiplication of orchids would be possible without sacrificing either the main apex or lateral buds of mother plants (Rao, 1977). However, shoot induction was not evident on any medium in the shoot and root tip explants from *in vivo* plants. The differential induction in the explants obtained from *in vitro* and *in vivo* grown seedlings could be due to differences in their physiological age. The importance of physiological age has been emphasized by Vajrabhaya and Vajrabhaya (1970) in leaf culture studies of *Rhynchostylis gigantea*. On Kn C medium + CM, axillary bud from *in vivo* plants of *D. chrysanthum* differentiated directly into shoots. CM is suitable for growth and differentiation due to presence of inositol and ammonium nitrogen (Shantz and Steward, 1964); and growth regulators (Letham, 1974; Van Standen and Drewes, 1975). Kn C medium with BA (0.1 and 0.5 mg 1^{-1}) proved suitable for shoot development in explants having inflorescence primordia obtained from *in vivo* grown plants of *D. chrysanthum*. Tanaka *et. al.*, (1988) also reported the essentiality of an exogenous supply of cytokinin for the induction of vegetative buds. Size of the inflorescence was critical for shoot induction. Whereas inflorescence primoridia 3mm × 3mm reverted to vegetative phase and developed shoots, inflorescence primordia 4mm × 3mm differentiated floral buds.

Except axillary buds which revealed single shoots; multiple shoots developed from shoot tip, root tip and inflorescence primordia (3mm × 3mm) on suitable medium. However, percentage respose of shoot tip, root tip explants to differentiation was less as compared to axillary bud and inflorescence primordia (3mm × 3mm) explants. Growth of shoots was better in the shoots which differentiated from inflorescence primorida and axillary bud explants. The growth of multiple shoot developed from inflorescence primordium explants cultured on 0.1 mg 1^{-1} BA supplemented medium was the best. Thus in the present investigation inflorescence primordia were by far the best source of explants for successful clonal propagation followed by axillary buds, shoot tips and root tip explants. In *P. spicerianum* shoot induction was not evident on any medium in the various culture explants obtained from *in vitro* and *in vivo* seedlings. Morel (1974), Cooper (1983), Huang (1988) also could not succeed in growing *Paphiopedilum* hybrid through meristem culture.

In the present study with cold preservation of protocorms of *D. chrysanthum* and *P. spicerianum* was attempted but success with protocorms of *P. spicerianum* could not be achieved due to loss of their regenerative potential. The cold treated protocorms of *D. chrysanthum* regenerated plantlets only when cultured on MKn C semi-solid medium. Kn C medium was unsuitable for regeneration of plantlets from cold preserved protocorms. The MKn C medium differed from Kn C medium in additional presence of vitamins, macroelements, microelements and in having lower concentrations of $Ca(NO_3)_2$, KH_2PO_4 and NH_4SO_4.

In the present study, *D. chrysanthum* protocorms preserved for a period upto 6 months at 4°C in Kn C liquid medium incorporated with 4 per cent sucrose, stayed green and when cultured on MKn C medium under normal cultural conditions, 60 per cent protocorms regenerated plantlets.

In the present investigation for the establishment of *in virtro* raised seedlings of *D. chrysanthum* and *P. spicerianum* in field conditions acclimatization of seedlings was essential because otherwise the seedlings do not survive. Acclimatization of the *in vitro* raised seedlings is important to minimize water stress (Mohammed and Vidaver, 1988). The seedlings of epiphytic species *D. chryasnthum* did not survive on potting medium suitable for the terrestrial species *P. spicerianum*. In case of *D. chrysanthum* maximum survival was observed with seedlings transferred to pots containing brick pieces, charcoal pieces, tree fern (*Cyathea* spp.) pieces, bark pieces, leaf mould and dry *Sphagnum* in 1 : 1 : 1 : 1 : 1 : 2 with and without green *Sphagnum* on top. In *P. spicerianum* maximum survival of seedling was evident when seedlings were transferred to pots containing established *Paphiopedilum* plants.

Based on the findings of the present investigations it may be suggested that for optimum seed germination, rapid multiplication and conservation of these species (*D. chrysanthum* and *P. spicerianum*) the following practices should be followed :

	D. chrysanthum	*P. spicerianum*
Seed germination and growth :		
Capsule age (days)	330	270
Light	Light	Dark for seed germination & light for chlorophyll differentiation and growth
Medium :	Knudson C	Thomale GD
(a) Carbohydrates	Sucrose (2%)	Sucrose (2%)
(b) Nitrogen source	Ca $(NO_3)_2$ (1.0 gl^{-1})	NH_4SO_4 (0.5 gl^{-1})
(c) Growth hormones	Kinetin induced better seedling growth	IAA & NAA increased seed germination
(d) Organic supplements	Not required	Not required
(e) pH	5.0	5.0
Clonal Propagation		
Medium	Mod. Knudson C (KnC + 0.1 mgl^{-1} BAP)	No success
Source of explant	Inflorescence primordium	—
Cold preservation of protocorms		
Temperature	4°C	No success
Time period	6 months	—
Medium for regeneration	Mod. Knudson C (KnC + 0.3 mgl^{-1} thiamin HCl + 0.3 mgl^{-1} niacin + 0.3 mgl^{-1} pyridoxin HCl)	—
Establishment of in *vitro* raised seedlings to field conditions :		
Acclimatization	To be done in humidity chamber	To be done in humidity chamber
Compost for establishment of seedlings in community pots	Brick pieces, bark pieces, charcoal pieces, tree fern (*Cyathea* spp.) pieces, leaf mould & dry *Sphagnum* in 1 : 1 : 1 : 1 : 1 : 2 with and without green *Sphagnum* on top	Pots containing established *Paphiopedilum* plants

Appendix

EFFECT OF GAMMA RADIATION ON PROTOCORMS

Introduction

Variations are common amongst orchids since they are usually cross pollinated plants. The natural gene pool has been utilized to good effect in the hybridization and selection of individuals carrying desirable genes (Vajrabhaya, 1977). Ionizing radiations are increasingly used for the improvement of crop plants including ornamentals. However, this technique has rarely been applied to orchids (Vajrabhaya, 1977), since orchids are slow growing plants which produce relatively few shoots per year (Harn, 1968; Vajrabhaya, 1977). Miyazaki (1968) studied radiosenitivity of orchid seeds and embryo. Harn (1968, 1970) considered irradiating seeds to be meaningless as orchids are cross pollinated and every seed in the same fruit is different genetically.

The new means of vegetative multiplication of orchids, i.e., the production of mericlonal protocorms, has opened the ways for the application of radiation and mutagen to orchid improvement (Harn, 1970).

Therefore, the present investigation was attempted to investigate radiation responses of the protocorms of *Dendrobium chrysotoxum* for future experimentation.

MATERIALS AND METHODS

Effects of gamma rays on seedling growth in protocorm cultures of *D. chrysotoxum* were investigated by irradiating one-month-old protocorms. Cobalt 60 source emitting gamma rays at a rate of 66.95 rad/second was used for the purpose.

Protocorms were raised by sowing approximately 30 seeds each in 25 culture tubes containing 20 ml Kn C medium. Methods for raising protocorms from seeds was as given in chapter 4 for *D. chrysanthum*. Out of 25 tubes/treatment, a set of five culture tubes with one month old protocorms were irradiated with different acute doses of gamma rays (i.e., 0.25 kR, 0.5 kR and 0.75 kR). A set of five tubes with unirradiated protocorms served as control. Subsequently, protocorms from each treatment were scooped out into petriplates, from which five protocorms were taken and subcultured in 50 ml conical flasks containing 30 ml of fresh Knudson C media incorporated with 50 g l^{-1} banana homogenate for seedling regeneration. Sixteen culture flasks were employed for each treatment as well as for the control. All the cultures were kept in continuous light (3,000 lux), at 24 ± 1°C for seedling development. At intervals of 60, 120, 180 and 210 days four cultured flasks were taken for observation from each treatment.

Table I : Effect of Gamma rays on seedling shoot height (mm) developed from irradiated protocorms

Treatments	*Days*			
(kR)	*60*	*120*	*180*	*210*
0.0	4.7 ± 0.17	8.3 ± 0.17	13.25 ± 0.2	14.7 ± 0.15
0.25	2.7 ± 0.17[a]	8.4 ± 0.16	13.4 ± 0.16	17.4 ± 0.2[a]
0.50	3.2 ± 0.2[a]	8.5 ± 0.17	13.6 ± 0.14	16.5 ± 0.17[ab]
0.75	3.7 ± 0.18[ab]	8.7 ± 0.17	13.8 ± 0.17	16.3 ± 0.13[ab]

Values are means (n=10) ± S.E.

a. Significantly different from control

b. Significantly different from 0.25 kR

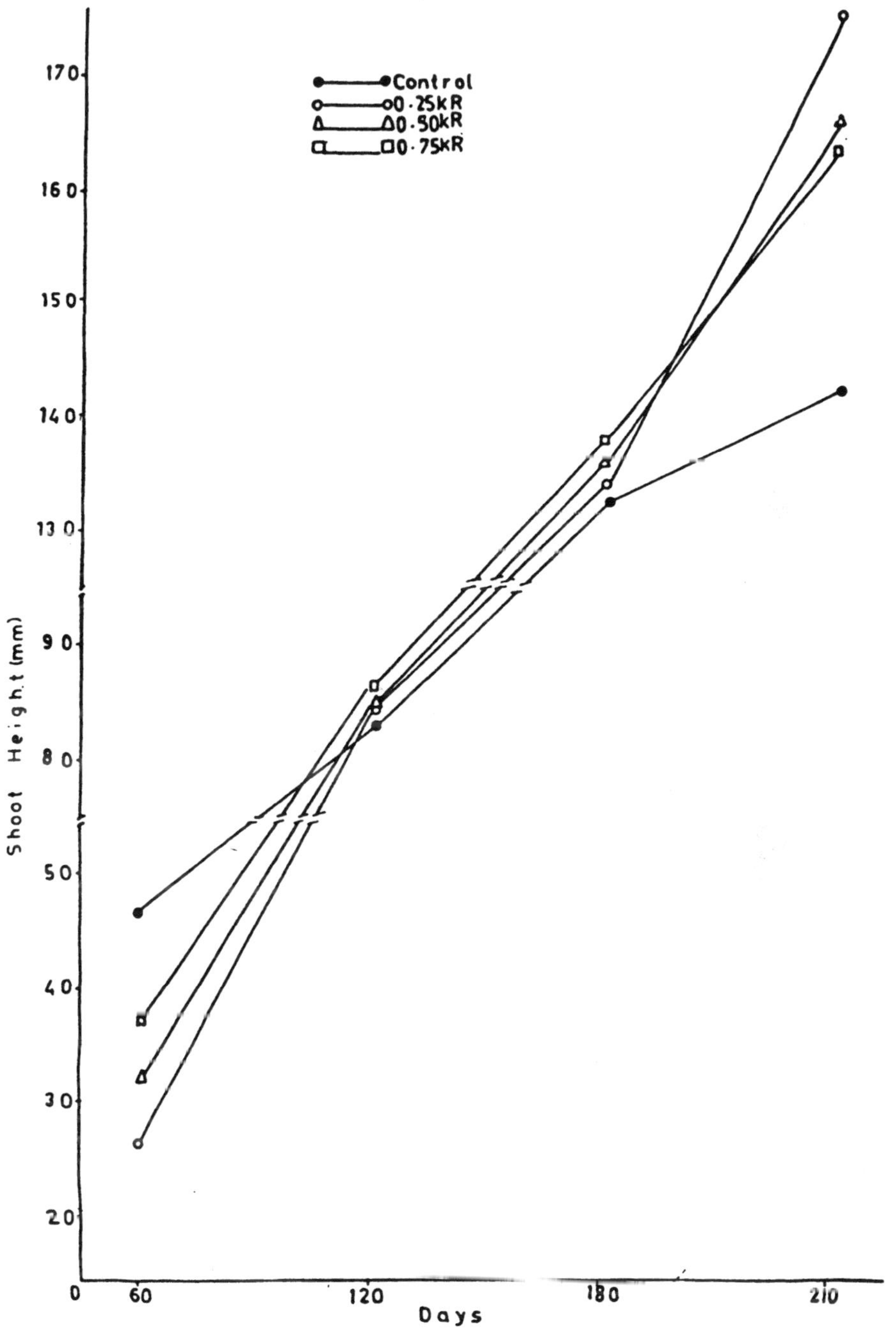

Fig. 1 : Radiation effect on shoot height (mm) *Dendrobium*

Plate 1a : Effect of radiation on seedling development from irradiated protocorms of *Dendrobium chrysotoxum*, after 60 days.

1. seedlings developing from unirradiated protocorms
2. seedlings developing from protocorms irradiated with 0.25 kR
3. seedlings developing from protocorms irradiated with 0.5 kR
4. seedlings developing from protocorms irradiated with 0.75 kR

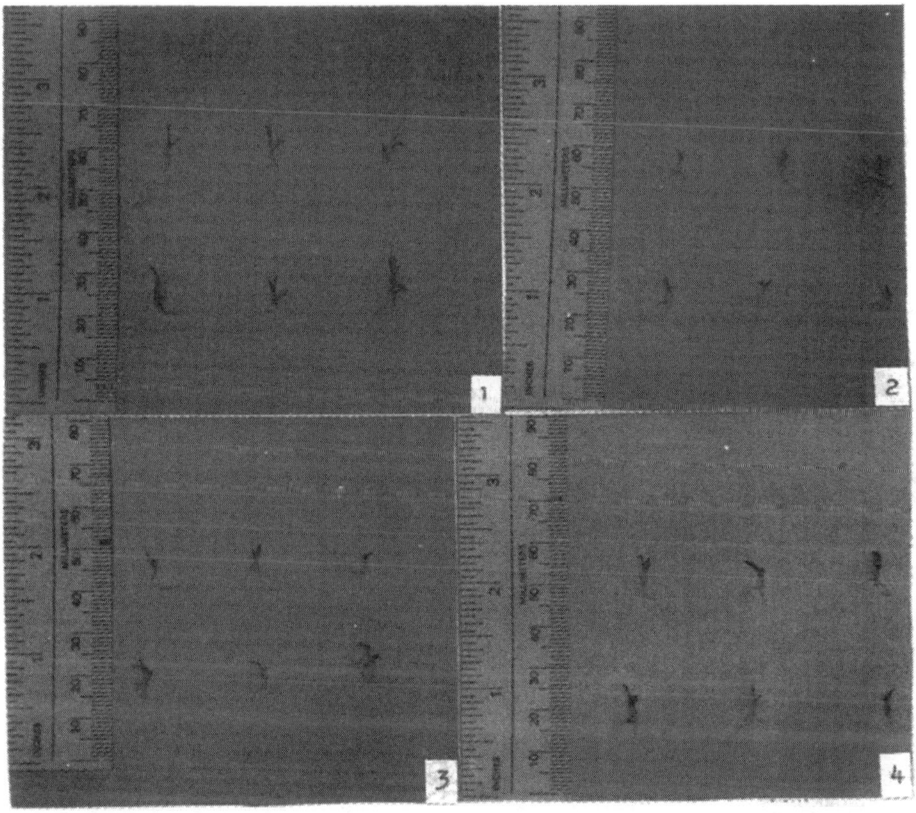
1
2
3
4

Differentiated seedlings from the four flasks of each treatment were placed together in one petriplate. Ten of the healthiest seedlings of each treatment were selected for the study of various growth parameters, viz., (a) shoot height, (b) leaf number, (c) leaf length and leaf breadth (d) root number, (e) root length, (f) roots with abnormal growth, (g) total length of root along with abnormal growth, and (h) fresh weight of the seedlings. Values of the parameters (c), (d), (f) and (h) was recorded as given in chapter 3.

Students 't' test was conducted for analysis of results.

RESULTS

Effects of gamma rays 0.25, 0.5 and 0.75 kR on seedling development from irradiated protocorms are given on Table I to VII and plate Ia. 1-4, Ib. 1-8.

Shoot Height

Control : The shoots developing from unirradiated protocorms measured 4.7 mm to 14.7 mm in 60-210 days old seedlings (Table I, Fig. 1).

Table II: Effect of gamma rays on seedling leaf number developed from irradiated protocorms

Treatments	*Days*			
(kR)	*60*	*120*	*180*	*210*
0.0	2.2 ± 0.1	2.5 ± 0.2	2.7 ± 0.2	3.0 ± 0.2
0.25	2.0 ± 0.0	2.6 ± 0.2	3.0 ± 0.1	3.4 ± 0.2
0.50	2.3 ± 0.1	2.8 ± 0.2	2.9 ± 0.1	3.4 ± 0.2
0.75	2.4 ± 0.2	2.9 ± 0.2	3.1 ± 0.2	3.5 ± 0.2

Values are means (n=10) ± S.E.

Table III : Effect of gamma rays on seedling leaf length and leaf breadth developed from irradiated protocorms.

Treatments	*Days*							
(kR)	*60*		*120*		*180*		*210*	
	L	*B*	*L*	*B*	*L*	*B*	*L*	*B*
0.0	3.6 ± 0.1	0.8 ± 0.0	5.6 ± 0.3	1.4 ± 0.0	8.4 ± 0.5	1.8 ± 0.0	14.2 ± 0.3	1.8 ± 0.0
0.25	2.4 ± [a] 0.1	0.8 ± 0.0	5.9 ± 0.3	1.6 ± 0.1	23.7 ± [a] 0.5	3.0 ± [a] 0.1	37.3 ± [a] 0.4	3.7 ± [a] 0.0
0.50	2.5 ± [a] 0.1	0.8 ± 0.0	6.2 ± 0.2	1.5 ± 0.1	22.4 ± [a] 0.5	2.9 ± [a] 0.0	28.1 ± [ab] 0.6	2.9 ± [ab] 0.0
0.75	3.2 ± [abc] 0.1	0.8 ± 0.0	6.5 ± [a] 0.3	1.5 ± 0.1	22.6 ± [a] 0.3	2.5 ± [abc] 0.0	28.3 ± [ab] 0.5	2.6 ± [abc] 0.0

Values are means (n=10) ± S.E.

a. Significantly different from control

b. Significantly different from 0.25 kR

c. Significantly different from 0.50 kR

L. Length (mm)

B. Breadth (mm)

Treated : The shoot height in 60 days old seedlings developed from protocorms irradiated with 0.25, 0.5 and 0.75 kR was significantly smaller than control (Table I, Fig. 1 and Plate Ia, 1-4). However, after 120 days the growth improved and height of the shoot at different radiation exposures compared well with the control shoots (Plate Ib. 1-4). But in 210 days old irradiated seedlings the shoot height decreased with the increasing exposure (Table I, Fig. 1).

Leaf Number

Control : The average (mean) leaf number in 60-210 days old control seedlings varied from 2.2 to 3.0 (Table II, Fig. 2).

Treated : Leaf number in irradiated seedlings did not differ from control (Table II, Fig. 2). The irradiated seedlings given different exposures did not differ from each other, in the number of leaves developed (Table II, Fig. 2).

Leaf Length and Breadth

Control : In control seedlings (60 days old) the leaf length and leaf breadth measured 3.6 and 0.8 mm respectively. The length and breadth of the leaves increased to 14.25 and 1.8 mm in 210 days old seedling (Table III).

Treated : 60 days old irradiated seedlings had significantly smaller leaf length compared to control. At 120 days the length of leaves given 0.25 and 0.5 kR radiation exposure eventhough resembled control the length of leaves given 0.75 kR was significantly more than the control (Table III). The length of the leaves in 180 and 210 days old irradiated seedlings were longer than leaves of the control seedlings. The radiation induced stimulation of leaf length decreased with the increasing exposure in 180 and 210 days old irradiated seedlings (Table III, Fig.3).

Table IV: Effect of gamma rays on seedling root number developed from irradiated protocorms

Treatments (kR)	Days 60	120	180	210
0.0	2.0 ± 0.1	2.3 ± 0.2	2.8 ± 0.3	2.9 ± 0.2
0.25	1.9 ± 0.3	4.2 ± 0.3 [a]	4.7 ± 0.3 [a]	5.1 ± 0.3 [a]
0.50	2.0 ± 0.1	3.3 ± 0.1 [ab]	3.6 ± 0.2 [ab]	4.0 ± 0.3 [ab]
0.75	2.7 ± 0.3	3.1 ± 0.2 [ab]	3.6 ± 0.2 [ab]	3.8 ± 0.2 [ab]

Values are means (n=10) ± S.E.

a. Significantly different from control

b. Significantly different from 0.25 kR

Breadth of the leaves in 60 and 120 days old irradiated seedlings did not differ from control. However, in 180 and 210 days old seedlings the leaves were significantly broader than the control leaves (Table III, Fig. 4.). The breadth of the leaves decreased with the increasing gamma ray exposure (Table III).

Plate Ib : Effect of radiation on seedling development from irradiated protocorms of *Dendrobium chrysotoxum*

1. seedlings developing from unirradiated protocorms, after 120 days
2. seedlings with abnormal growth in the roots, developing from protocorms irradiated with 0.25 kR, after 120 days
3. seedlings with abnormal growth in the roots, developing from protocorms irradiated with 0.5 kR. after 120 days
4. seedlings developing from protocorms irradiated with 0.75 kR, after 120 days
5. seedling developing from 0.25 kR irradiated protocorms, after 180 days
6. seedling developing from 0.5 kR irradiated protocorms, after 180 days
7. seedling developing from 0.25 kR irradiated protocorms, after 210 days
8. seedling developing from 0.5 kR irradiated protocorms, after 210 days
9. transverse section of control root x 800
10. transverse section of irradiated root through abnormal swelling x 397

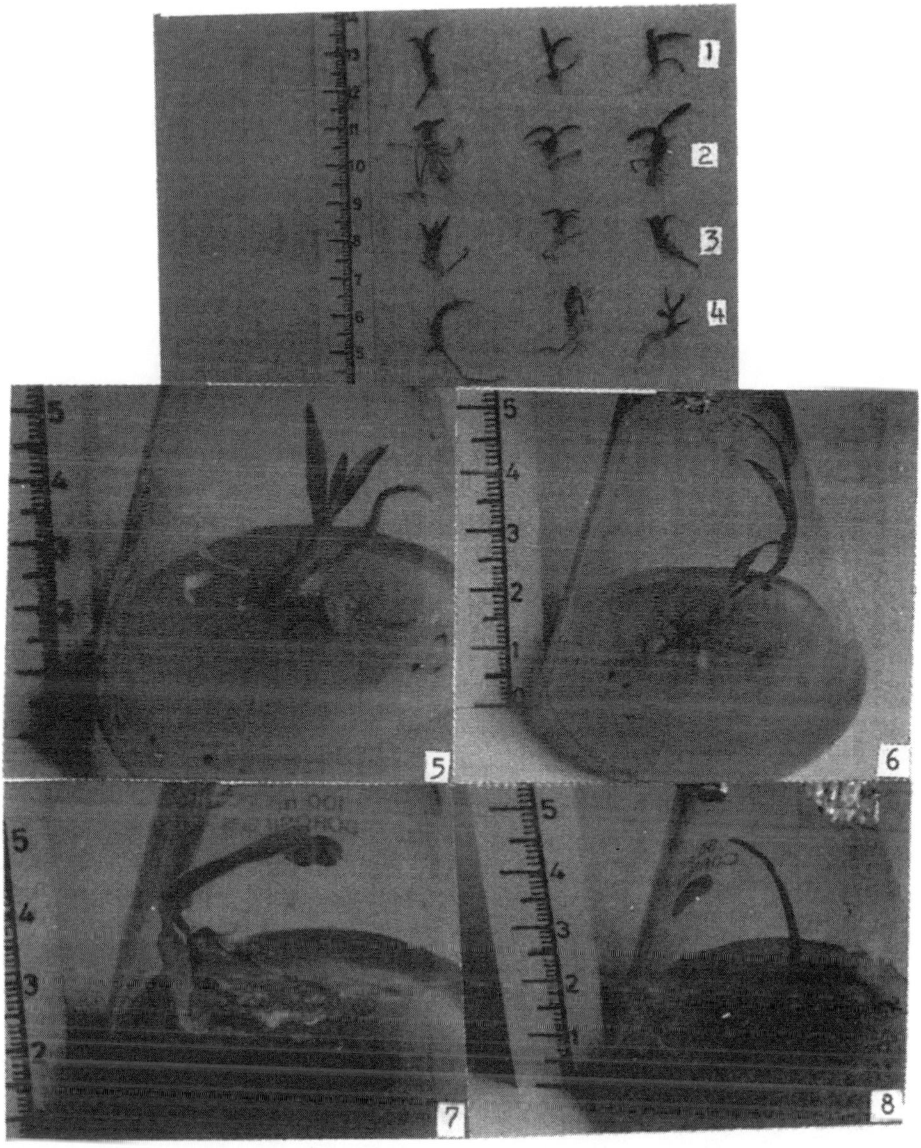

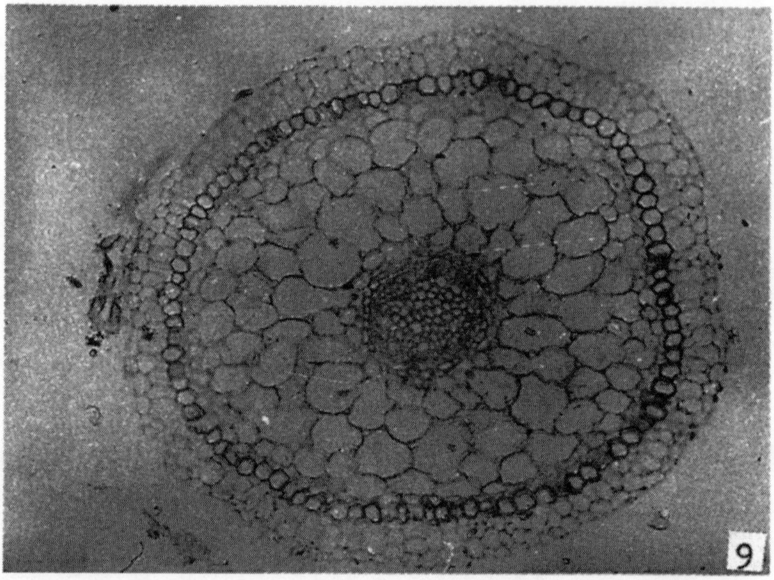

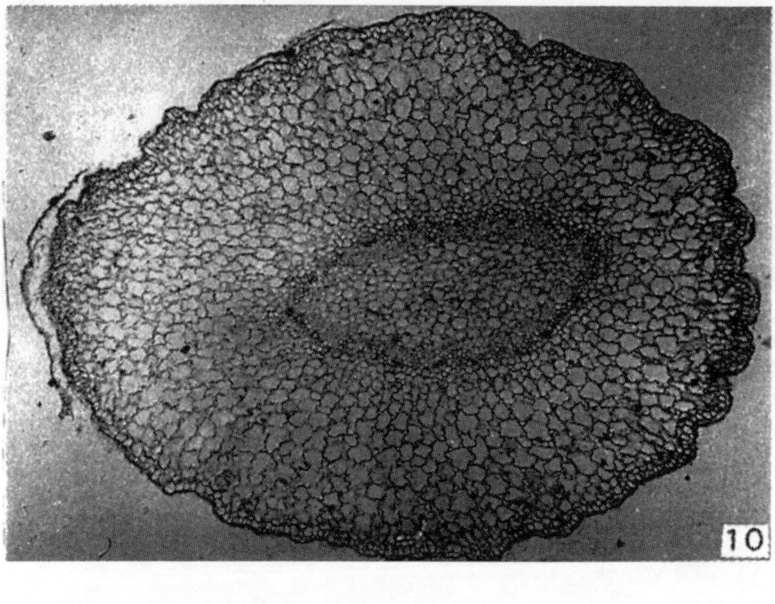

Plate 1 b

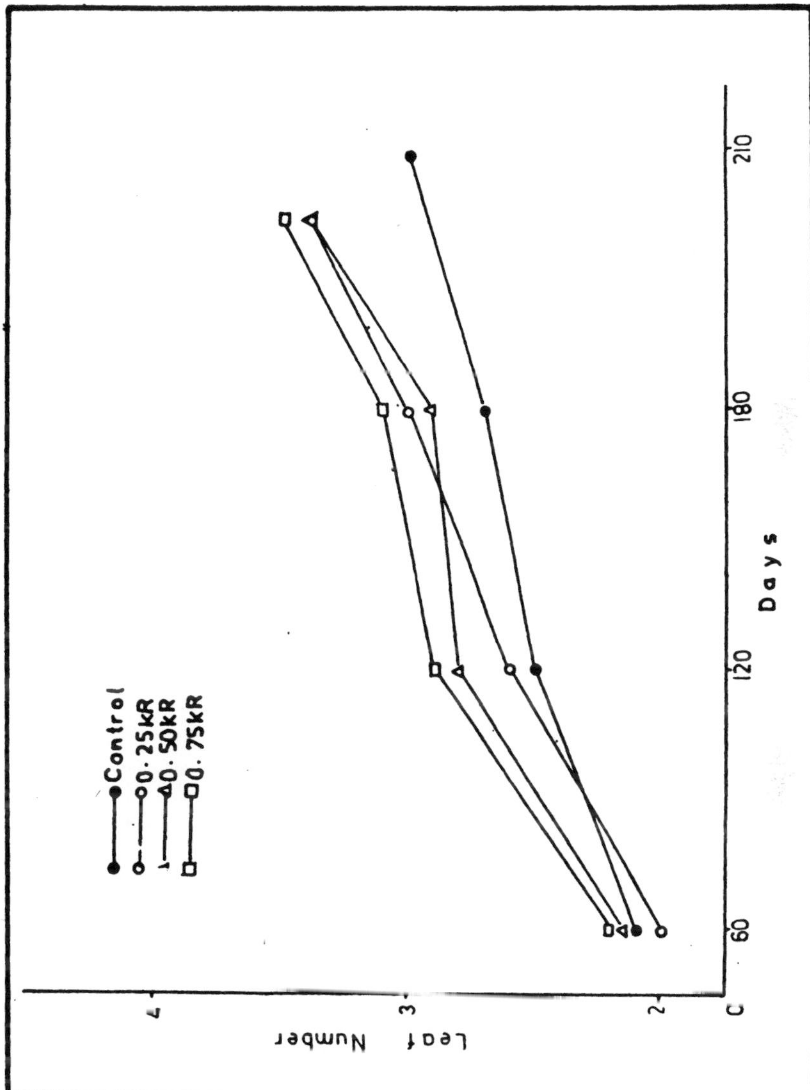

Fig. 2 : Radiation effect on leaf number *Dendrobium*

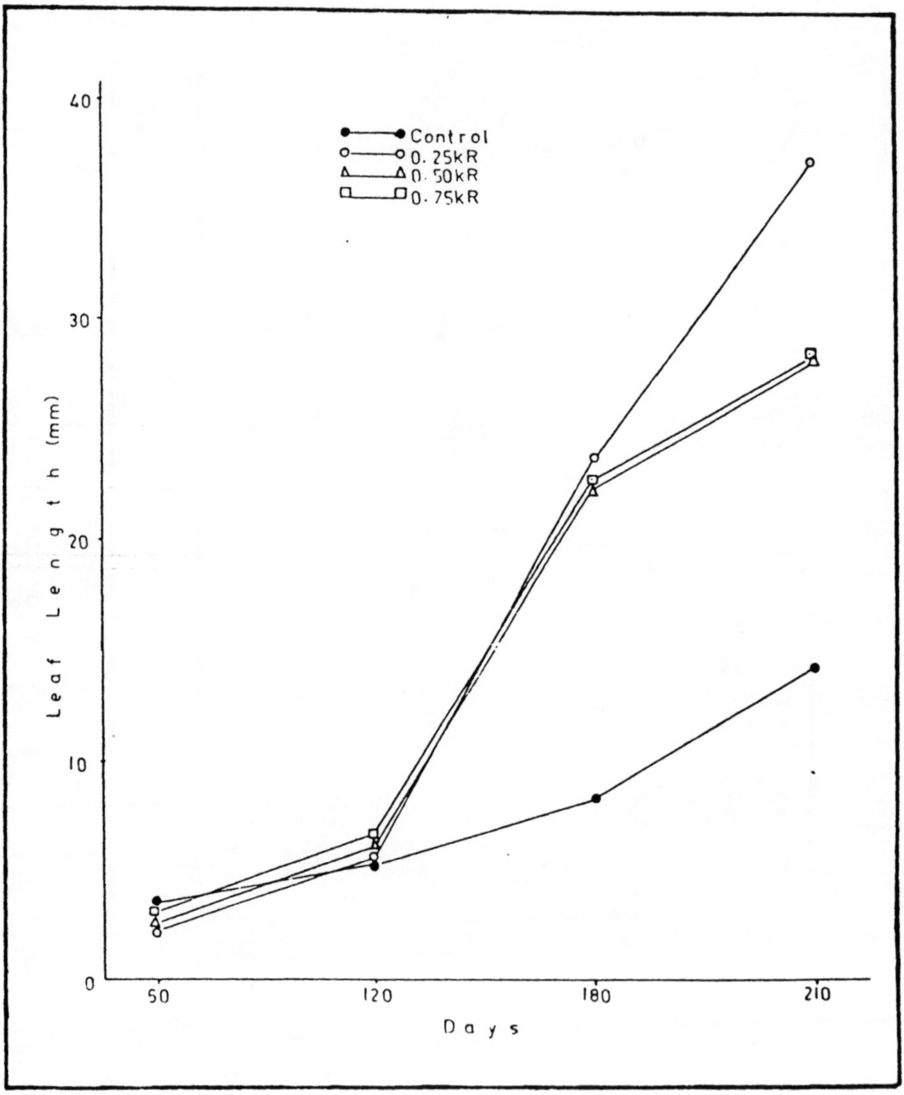

Fig. 3 : Radiation effect on leaf length (mm) *Dendrobium*

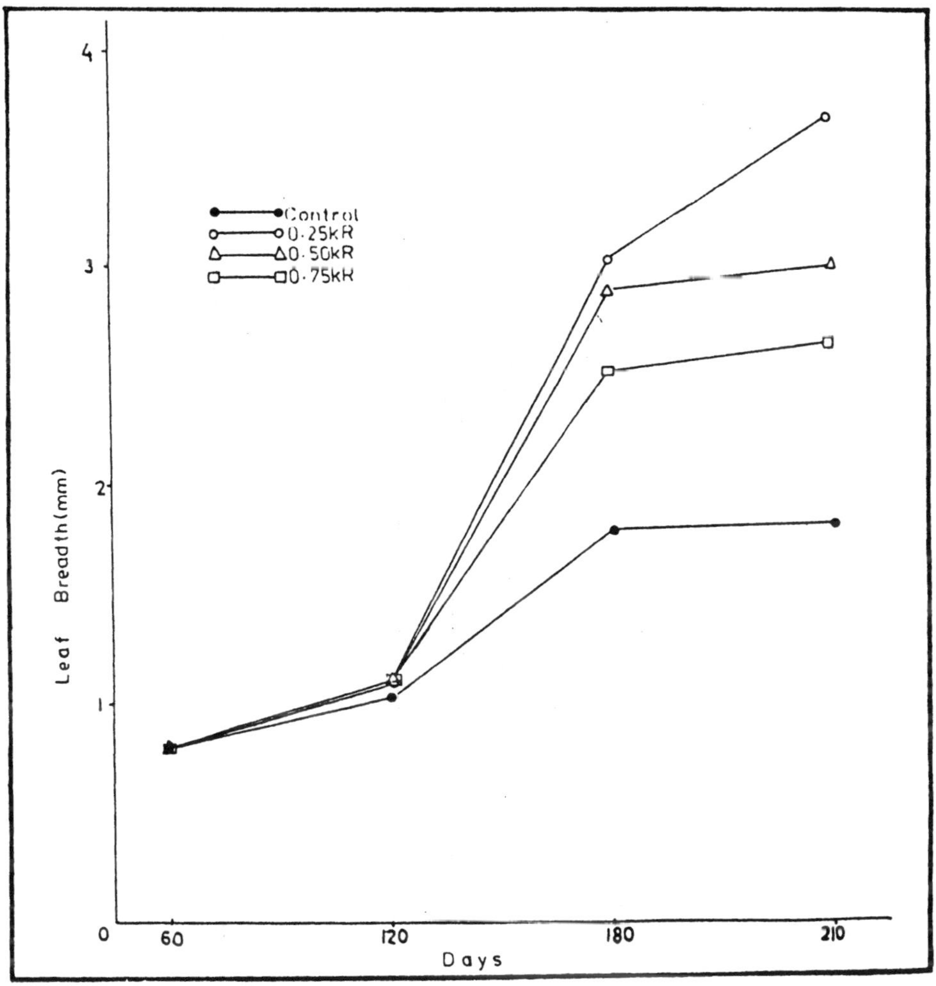

Fig. 4 : Radiation effect on leaf breath (mm) *Dendrobium*

Root Number

Control : The average (mean) number of roots in seedlings developed from unirradiated protocorms varied between 2.0-2.9 in 60-210 days old seedlings (Table IV).

Treated : To begin with, the number of roots in irradiated seedlings did not differ from control (Table IV).

Amongst the 60 days old irradiated seedlings, maximum number of roots were observed in seedlings given 0.75 kR radiation exposure (Table IV). In 120-210 days old irradiated seedlings the number of roots was significantly higher than control (Table IV). The number of roots developed however decreased with increasing exposure (Table IV, Fig. 5).

Table V: Effect of gamma rays on seedling root length (mm) developed from irradiated protocorms

Treatments	*Days*			
(kR)	*60*	*120*	*180*	*210*
0.0	2.9 ± 0.2	7.0 ± 0.2	7.3 ± 0.2	9.7 ± 0.4
0.25	2.8 ± 0.1	6.7 ± 0.4	7.6 ± 0.3	23.5 ± 0.2 [a]
0.50	3.4 ± 0.6	7.3 ± 0.4	7.8 ± 0.2	25.6 ± 0.3 [ab]
0.75	3.6 ± 0.1 [a]	19.0 ± 1.7 [abc]	22.3 ± 1.3 [abc]	37.7 ± 0.6 [abc]

Values are means (n=10) ± S.E.

a. Significantly different from control

b. Significantly different from 0.25 kR

c. Significantly different from 0.50 kR

Table VI: Root length (mm) along with abnormal growth of seedling developed from irradiated protocorms.

Treatments	*Days*			
(kR)	*60*	*120*	*180*	*210*
0.0				
0.25		6.3 ± 0.5	12.3 ± 0.5	39.7 ± 1.2
0.50		6.0 ± 0.2	9.2 ± 0.2 [b]	21.9 ± 0.9 [b]
0.75				

Values are means (n=10) ± S.E.

a. Significantly different from 0.25 kR

Root Length

Control : The length of the roots developed from differentiated control shoots ranged between 2.9 to 9.6 mm in 60-210 days old seedlings (Table V).

Treated : Length of the roots in 60 days old seedlings given 0.25 and 0.5 kR radiation did not differ from control (Table V). In 60 and 210 days old seedlings given 0.75 kR radiation the roots were significantly longer in comparison to control (Table V). Unlike this the roots of 120 and 180 days old seedlings exposed to 0.25 and 0.5 kR did not differ from control in size (Table V). However, in 210 days old seedlings irradiated with 0.25-0.75 kR gamma ray exposure the roots were significantly longer in comparison to control (Table V). Usually the size of the root increased with the increasing exposure (Table V, Fig. 6).

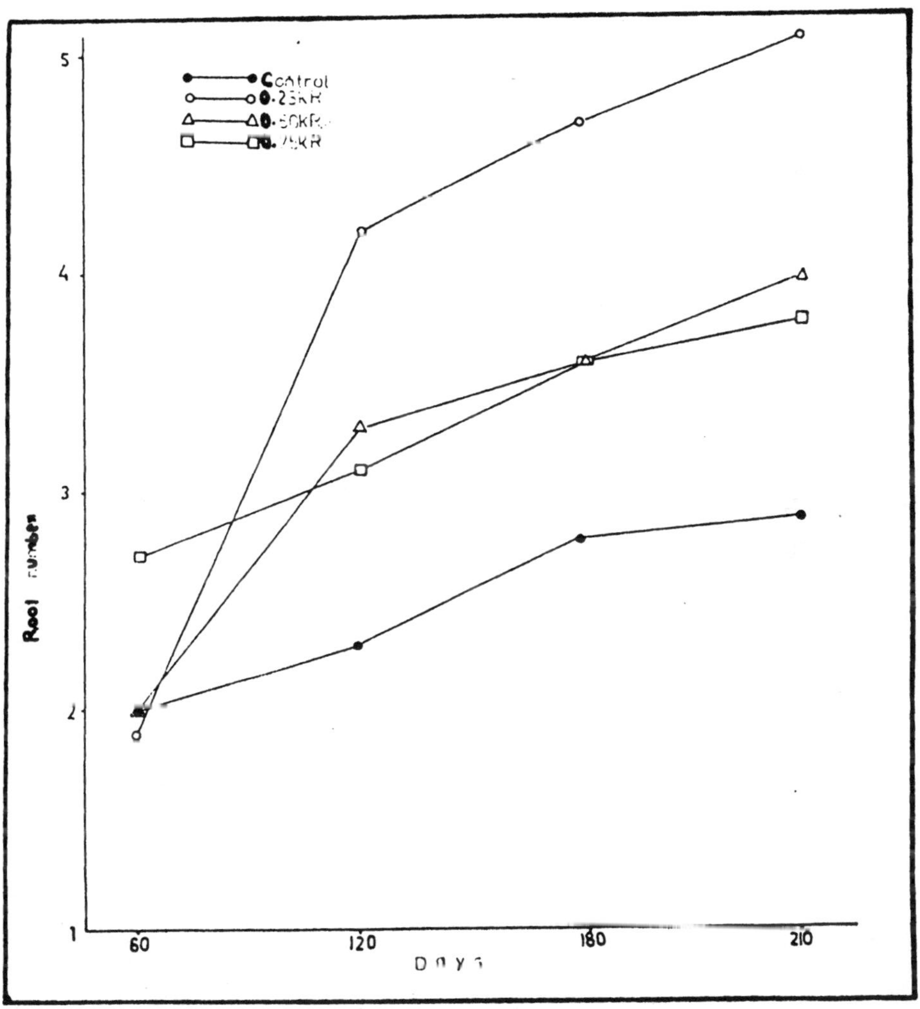

Fig. 5 : Radiation effect on root number *Dendrobium*

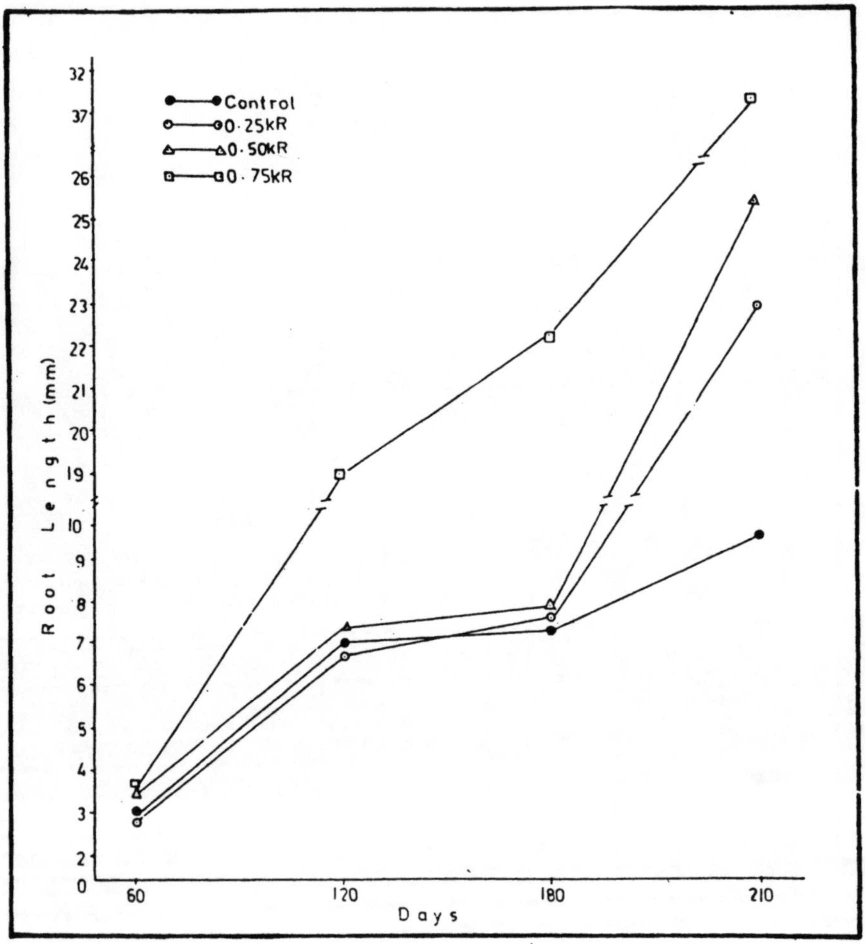

Fig. 6 : Radiation effect on root length (mm) *Dendrobium*

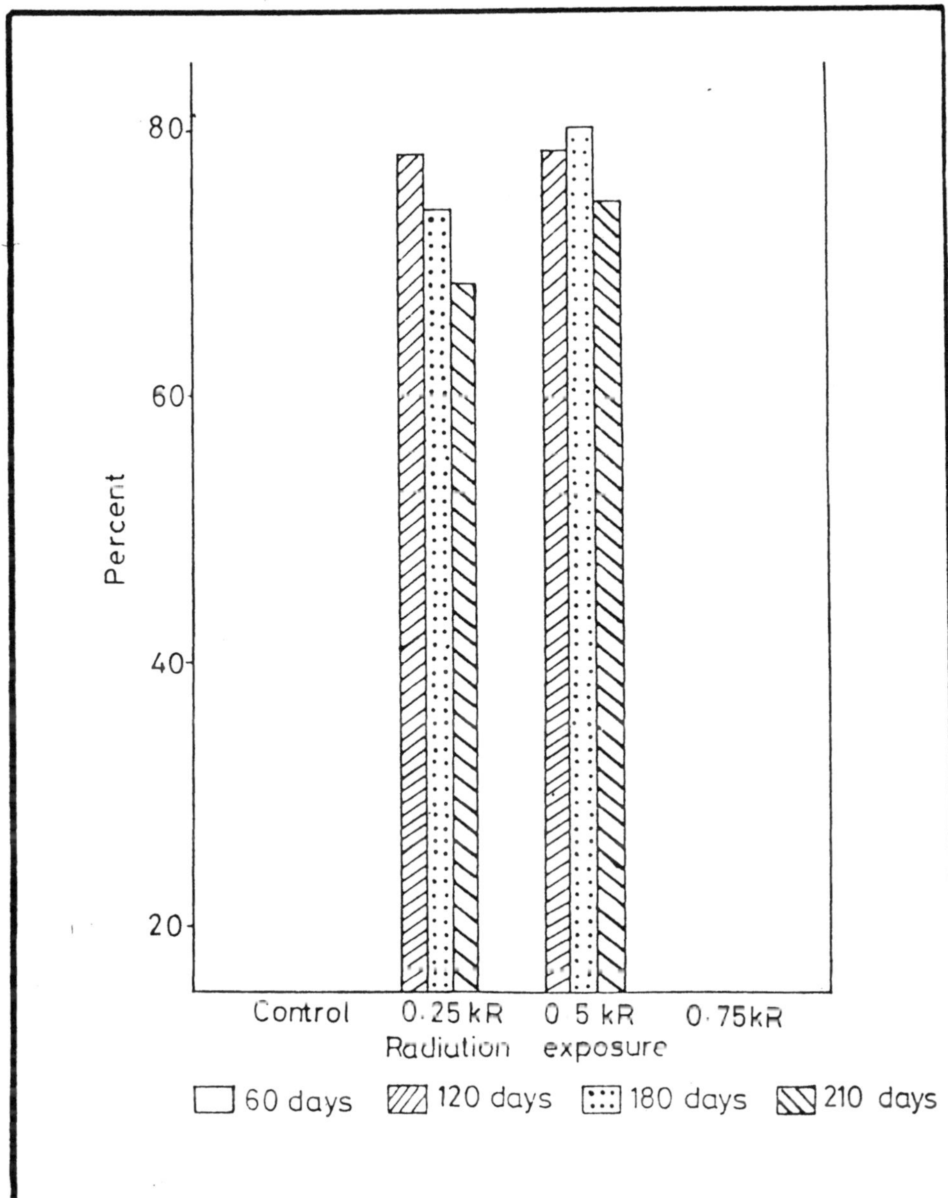

Fig. 7 : Radiation effect on root abnormality *Dendrobium*

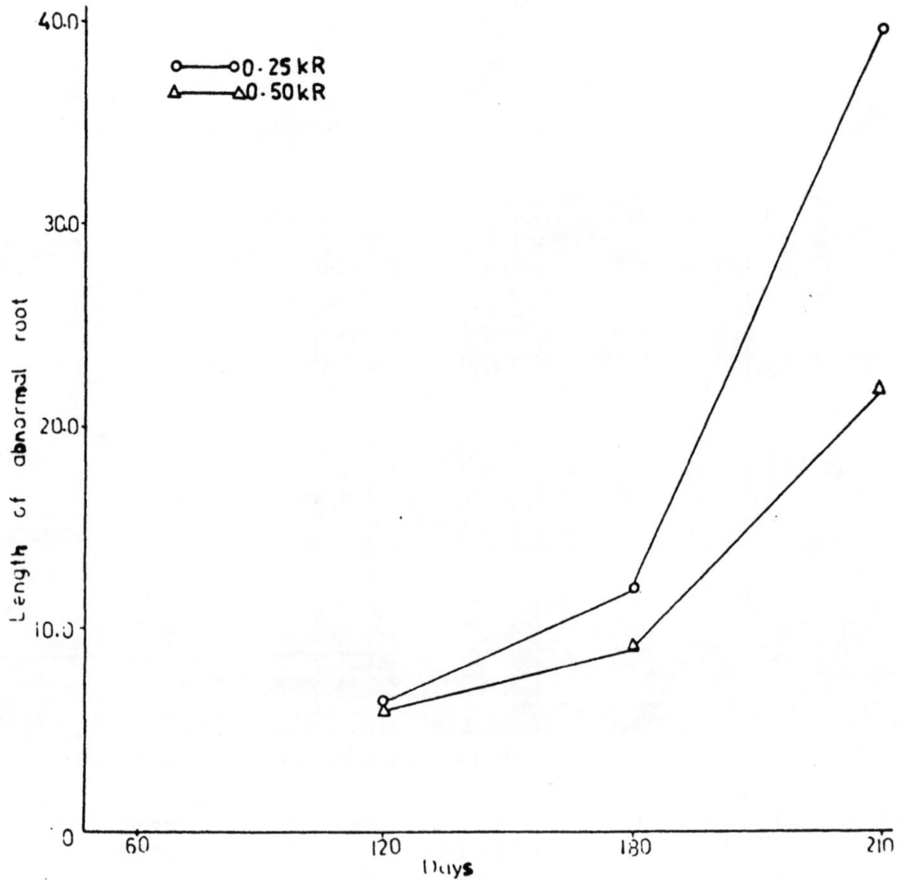

Fig. 8 : Radiation effect on length of abnormal root (mm) *Dendrobium*

Roots with Abnormality

Control : No abnormality was observed in roots of control seedlings (Table VI).

Treated : Eventhough roots of 60 days old irradiated seedlings exhibited no abnormality, abnormal growth developed on the roots of 120 days old irradiated seedlings given 0.25 and 0.5 kR radiation. In irradiated seedlings some of the roots revealed swelling of the root tip region (Plate Ib. 2-3). In 180 days old irradiated seedlings these enlarged and became somewhat triangular in shape (Plate Ib. 5-6) and in 210 days seedlings roots developed from these swollen portion (Plate Ib. 7-8). The abnormal growth were bigger in roots of seedlings given 0.25 kR, in comparison to the seedlings developed in roots given 0.5 kR (Plate Ib. 5-6). Transverse section of the swellings revealed excessive enlargement and proliferation of cortical cells (Plate Ib. 9-10).

Compared to seedlings irradiated with 0.5 kR the number of roots with abnormal growth was higher in seedlings given 0.25 kR (Fig. 7).

The roots exhibiting abnormal growth were longer in seedlings given 0.25 kR compared to such roots of seedlings given 0.5 kR (Table VI, Fig. 8).

Number and length of roots developed from abnormal growth of roots in 0.25 kR irradiated seedlings was higher than in the roots of seedlings given 0.5 kR (Plate Ib, 7-8). Multiple adventitious roots developed from the abnormal root growth of 0.25 kR dose rate (Plate Ib. 7).

Fresh Weight of Seedlings

Control : Fresh weight of seedlings developed from unirradiated protocorms varied from 10.1-26.2 mg in 60-210 days old seedings (Table VII).

Table VII: Effect of gamma rays on seedlings fresh weight (mg) developed from irradiated Protocorms.

Treatments	*Days*			
(kR)	*60*	*120*	*180*	*210*
0.0	10.1 ± 0.9	20.7 ± 1.2	23.4 ± 1.4	26.2 ± 1.5
0.25	7.0 ± 0.5^{a}	49.2 ± 2.8^{a}	84.5 ± 3.0^{a}	104.0 ± 7.6^{a}
0.50	9.8 ± 0.4^{a}	46.0 ± 1.9^{a}	56.6 ± 1.7^{ab}	111.2 ± 5.1^{ab}
0.75	10.7 ± 0.7^{b}	32.0 ± 2.0^{abc}	34.7 ± 1.9^{abc}	45.7 ± 2.4^{abc}

Values are means (n=10) ± S.E.

a. Significantly different from control

b. Significantly different from 0.25 kR

c. Significantly different from 0.50 kR

Treated : Fresh weight of 60 days old irradiated seedlings (0.25 and 0.5 kR) was significantly less than control. The fresh weight of seedlings given 0.75 kR did not differ from control (Table VII). However, 120 days onwards the seedlings from 0.25-0.75 kR radiation revealed higher fresh weight compared to control, (Table VII, Fig, 7.9). The stimulation of fresh weight over control was maximum

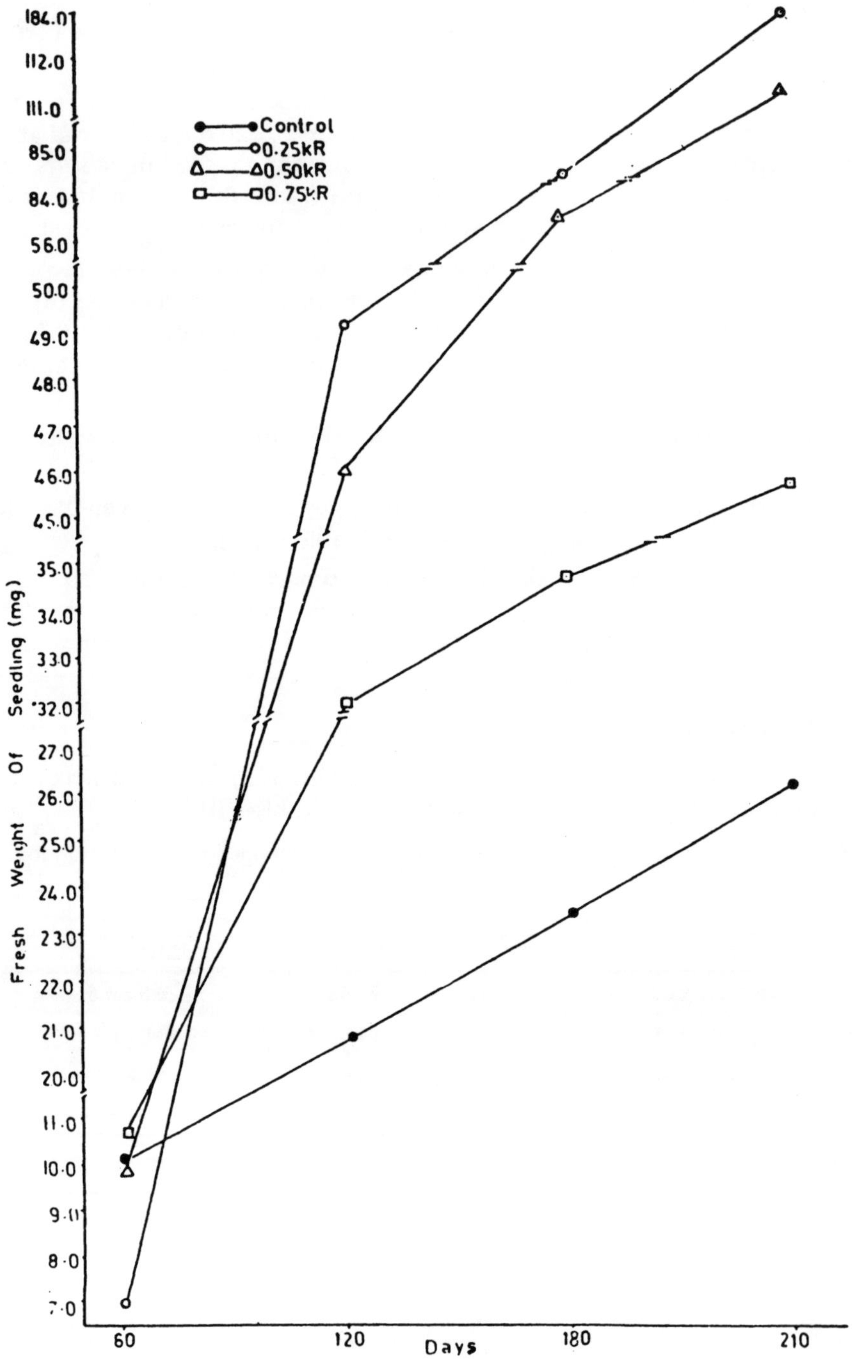

Fig. 9 : Radiation effect on fresh weight of seedlings (mg) *Dendrobium*

at 0.25 kR (Table VII). The degree stimulation decreased with the increasing radiation exposure (Table VII).

DISCUSSION

During the present investigation compared to control seedlings the shoot height was significantly inhibited in 60 days old irradiated seedlings. Radiation-induced reduction of seedling height is reported in the literature and could be due to the damage of apical meristem (Iqbal, 1969; Patel and Shah, 1974; Chauhan, 1978), slow rate of meristematic activity and/or arresting of mitotic cycle (Rao and Rao, 1983). The shoot height in 120 and 180 days old irradiated seedlings did not differ from control. However in 210 days old irradiated seedlings there was stimulation of shoot height over control. The above findings therefore suggest a recovery from radiation damage. Recovery from radiation damage with the passage of time is reported in literature (Iqbal, 1970; Gunckel and Sparrow, 1961). Radiation-induced stimulation of growth and differentiation has also been reported in tobacco tissue culture (Degani and Pickholz, 1973), in orange ovular culture (Spiegel and Kochba, 1973).

In case of root with normal growth, the root length of seedlings developed from irradiated protocorms usually increased with the increasing exposure and age of the seedlings. Radiation-induced increase of root length is reported in *Pinus rigida* (Mergen and Johansen, 1964). *Yucca brevifolia* (Johnstone and Klepinger, 1967), *Phaseolus torosus* (Sinha and Roy, 1984).

The control roots and roots given 0.75 kR irradiation did not develop any abnormality; on the other hand, roots of seedlings developing from protocorms irradiated with 0.25 and 0.5 kR irradiation formed normal roots as well as roots with abnormal growth on the root tips. Abnormal growth of roots in seedlings developed from treated protocorms by lower radiation exposures, is due to the active cell proliferation towards the root tips. Jain *et al.*, (1984) reported that in the anther culture of *Datura innoxia* low exposure (0.2 kR) of gamma rays was conducive for active cell proliferation resulting in exuberant growth of callus.

Seedlings developing from irradiated protocorms had higher fresh weight than control. Radiation-induced stimulation of the fresh weight of seedlings is reported in *Phaseolus vulgaris* (Bajaj *et al.*, 1970) and *Vigna mungo* (Rao and Rao, 1983). In the present investigation radiation induced increase in fresh weight of irradiated seedlings over control. However, the stimulation of seedling fresh weight decreased with the increasing exposure.

References

Bajaj, V.P.S., A.W. Saettler, and M.W. Adams. 1970. Gamma irradiation studies on seeds, seedlings and callus tissue cultures of *Phaseolus vulgaris* L. *Rad. Bot.*, 10 : 119-24.

Chauhan, Y.S. 1978. Gamma-ray induced variation in the development of S. *khasianum* Clarke. *J. Indian Bot. Soc.*, 57 : 347-52.

Degani, N., and D. Pickholz. 1973. Direct and indirect effect of gamma irradiation on the differentiation of tobacco tissue culture. *Rad. Bot.*, 13 (1-6) : 381.

Harn, C. 1968. Propagation of orchids and mutation breeding. *Amer. Orchid Soc. Bull.*, 37 : 1066-68.

Harn, C. 1970. Radiosensitivity of *Cymbidium* protocorms. *Amer. Orchid Soc. Bull.*, 39 : 499-505.

Iqbal, J. 1969. Radiation induced growth abnormalities in vegetative shoot apices of *Capsicum annuum* L. in relation to cellular damage. *Rad. Bot.*, 9 : 491-99.

Iqbal, J.1970. Recovery from cellular damage in vegetative shoot apices of *Capsium annuum* L. after acute gamma irradiation. *Radiat. Bot.*, 10 : 337-43.

Jain, R.K., N. Maherchandani, D.R. Sharma, and J.B. Chowdhury. 1984. Effect of gamma radiations and gibberellic acid on growth and shoot regeneration in callus cultures of *Datura innoxia. Curr. Sci.*, Vol. 53 (13). pp. 700-701.

Johnstone and Klepinger. 1967. The effects of gamma radiation on germination and seedling development of *Yucca brevifolia* Engelm. *Rad. Bot.*, 7 : 385-88.

Mergen, F., and T.S. Johansen. 1964. Effect of ionizing radiation on seed germination and seedling growth of *Pinus rigida* Mill. *Rad. Bot.*, 4 : 417-27.

Miyazaki, S. 1968. On the germination and growth of the orchid seeds irradiated by gamma rays. *Agric. Bull.* Saga Univ., 43 : 101-16.

Patel, J.D., and J.T. Shah. 1974. Effect of gamma irradiation on seed germination and organization of shoot apex in *Solanum melongena* and *Capsicum annuum. Phytomorph.*, 24 : 174-180.

Rao, S.R., and Rao, D. 1983. Gamma ray induced physiological variabilities in black gram (*Vigna mungo*) L.) *Indian J. Bot.*, 6 (2) : 40-44.

Sinha, S.S.N., and H. Roy. 1984. Study of radiosensivity towards the growth of the seedlings in some *Phaseolus* species. In : R.P. Sinha (ed.) *Recent Trend in Bot. Researches*, pp. 181-85, Mulidhar Press, Musallahpur, Patna, India.

Spiegal-Roy, P., and J. Kochba. 1973. Stimulation of differentiation in orange (*Citrus sinensis*), ovular callus in relation to irradiation of the mediä. *Rad. Bot.*, 13 (1-6) : 97.

Vajrabhaya, T. 1977. Variation in clonal propagation. 178-207. In : Arditti, J. (ed.) *Orchid Biology : Reviews and Perspectives*, 1. Cornell Univ. Press, Ithaca.

Bibliography

Afzelius, K. 1916. Zur Embryosackentwicklung der Orchideen. *Svensk Bot. Tidskr.* 10 : 183-227.

Alvensleben, R.V. 1980. Marketing of tropical orchids from Asean countries in Western Europe. In: *Proc. 3rd ASEAN Orchid Congress,* Kuala Lumpur. 176-202.

Anonymous. 1890. L'hybridation des orchidées. *Jour. Orch.* 1 (17) : 267-269, 1 (18) : 283-286 (Attributed to L. Linden, Belgium).

Anonymous. 1923 a. Nonsymbiotic germination of Orchid seeds. *Orch. Rev.* 31 (355) : 13-14.

Anonymous. 1923 b. Seeds of the British Dactylorchids. *Orch. Rev.* 31 (358) : 100.

Anonymous. 1924. Germination of Orchid seed. *Orch. Rev.* 32 (373) : 193-194.

Anonymous. 1925 a. Germination of Orchid seed. *Orch. Rev.* 32 (382) : 98.

Anonymous. 1925 b. Germination of Orchid seed. *Orch. Rev.* 33 (386) : 225-226.

Anonymous. 1925 c. Symbiotic germination of Orchid seeds. *Orch. Rev.* 33 (386) : 249.

Arditti, J. 1961. *Cycnoches ventricosum* Batem. Var. *Chlorochilon* (Klotzsch) P.H. Allen *Comb. nov. Ceiba* 9(2) : 11-22.

Arditti, J. 1964. Culture media for Orchids. *Orch. Soc. Southern California Rev.* 6 (5): 13-15.

Arditti, J. 1967. Factors affecting the germination of orchid seeds. *Bot. Rev.* 33 : 1-97.

Arditti, J. 1977. Clonal propagation of orchids by means of tissue culture - a manual. pp. 202-293. In : J. Arditti (ed.), *Orchid biology : reviews and perspectives.* I. Cornell University Press, Ithaca, N. York.

Arditti, J. 1979. Aspects of the physiology of orchids. *Adv. Bot. Res.* 7 : 421-655.

Arditti, J., and R.F. Bils. 1965. The germination of an Orchid seed. *Orch. Soc. Southern California Rev.* 7 (5) : 5-6.

Arditti, J., and R. Ernst. 1984. Physiology of germinating orchid seeds. In : *Orchid biology : reviews and perspectives,* III, pp. 177-222. Cornell University Press, N. York.

Arditti, J., P.L. Healey, and R. Ernst. 1972. The role of mycorrhiza in nutrient uptake of Orchids. II. Extracellular hydrolysis of oligosaccharides by asymbiotic seedlings. *Amer. Orchid Soc. Bull.* 41 : 503-510.

Arditti, J., Michaud, J.D., and A.P. Oliva. 1981. Seed germination of North American orchids. I. Native California and related species of *Calypso, Epipactis, Goodyera, Piperia* and *Platanthera. Bot. Gaz.* 142 (4) : 442-453.

Arditti, J., M.A. Clements, G. Fast, G. Hadley, G. Nishimura, and R. Ernst. 1982. Orchid seed germination and seedling culture-A manual. In : *Orchid Biol-*

ogy : reviews and perspectives, Vol. II (ed. J. Arditti), pp. 244-370. Cornell Univ. Press, Ithaca and London.

Arora, Y. K., and A. Mukherjee. 1983. Ornamental orchids of North Eastern India. *Tech. Bull.* No. 5 : ICAR, Shillong.

Attawar, M. 1992. Micropropagation of orchids. *J. Orchid Soc. India*, 6 (1, 2) : 5-9.

Ballion, G., and M. Ballion. 1924. The non-symbiotic germination of Orchid seed in Belgium. *Orch. Rev.* 32 (376) : 305-308.

Ballion, G., and M. Ballion. 1928. Asymbiotic germination of Orchid seed. *Orch. Rev.* 36 (418) : 103-112.

Ballard, W.W. 1987. Sterile propagation of *Cypripedium reginae* from seeds. *Amer. Orchid Soc. Bull.*, 59 (9) : 935-946.

Bajaj, Y.P.S. 1979. Technology and prospects of cryopreservation of germplasm. *Euphytica*, 28 : 267-285.

Bernard, N. 1899. Sur la germination du Neottia nidusavis. *Compt. Rend. Acad. Sci.* Paris 128 : 1253-1255.

Bernard, N. 1903. La germination de orchidées. *Compt. Rend. Acad. Sci.*, 137 : 483-485. Paris.

Bernard, N. 1904. Recherches expérimentales sur les Orchidées. *Rev. Gén. Bot.* 16: 405-451; 458-478.

Bernard, N. 1909. Lévolution dans la symbiose, les orchidées et leurs champignons commensaux. *Ann. Sci. Nat. Bot.*, Series 9, No. 9 : 1-196.

Bertsch, W. 1966. A new frontier : orchid propagation by meristem culture. In : De Garmo, L.R. (ed.) : *Proc. 5th World Orchid Conf.*, pp. 225-229. Long Beach.

Bhojwani, S.S. 1981. A tissue culture method for propagation and low temperature storage of *Trifolium repens* genotypes. *Physiol. Plant.* 52 : 187-190.

Bhojwani, S.S., and M.K. Razdan. 1983. Plant Tissue Culture : Theory and Practice. Developments in Crop Science. 5. Elsevier, New York, viii + 502 pp.

Blowers, J.W. 1958. Gibberellin for Orchids. *Orch. Rev.* 66 (780) : 128-130.

Borriss, H. 1969. Samenvermechrung und Anzucht europaischer Erdorchideen. pp. 74-78. In : *Proc. 2nd Eur. Orchid Congress.*

Bose, T.K., and S.K. Bhattacharjee. 1980. Orchids of India. Naya Prakash, Calcutta.

Bose, T.K., and T.P. Mukherjee. 1976. Effect of growth substances on growth, protocorm and plantlet formation in seedlings of *Cymbidium giganteum. Orchid Rev.* 84 : 107-108.

Bouriquet, G. 1947. Sur la germination des graines de *Vanillier (Vanilla planifolia* And.). *Agron-Trop.* 2 (3-4) : 150-164.

Brainerd, K.E., and L.H. Fuchiagami. 1981. Acclimatization of aseptically cultured apple plants to low relative humidity. *J. Am. Soc. Hortic. Sci.* 106 : 515-518.

Breddy, N. C. 1953. Observations on the raising orchids by asymbiotic cultures. *Amer. Orch. Soc. Bull.* 22 (1) : 12-17.

Breddy, N. C. 1991. Orchid mycorrhiza and symbiotic raising techniques. *Am. Orchid Soc. Bull.* 60 : 556-569.

Burgeff, H. 1909. Die Wurzelpilze der Orchideen, ihre kultur und ihre Leben in der Pflanze. G. Fischer Verlag, Jena, 220 pp.

Burgeff, H. 1934. Pflanzliche Avitaminose und ihre Behebung durch Vitamin Zufuhr. *Ber. Deut. Bot. Ges.* 52 : 384-390.

Burgeff, H. 1936. Samenkeimung der Orchideen. G. Fischer Verlag, Jena, 312 pp.

Burgeff, H. 1954. Samenkeimung und Kultur Europaischer Erdorchideennebst Versuchen zu ihrer Verbreitung. G. Fischer, Stuttgart. 48 pp.

Burgeff, H. 1959. Mycorrhiza of Orchids. In : *The orchids*, ed. by C. L. Withner, The Ronald Press Co., New York, pp. 361-395.

Burström, H. 1948. Observation on the influence of galactose on Wheat roots. *Physiol. Plant.* 1 : 209-215.

Carlson, M. C. 1935. The germination of the seed and development of the seedlings of *Calopogon pulchillus* (Sw.) R. Br. *Trans. Illinois State Acad. Sci.* 28 (2) : 85-86.

Carlson, M. C. 1940. Formation of the seed of *Cypripedium parviflorum. Bot. Gaz.* 102 (2) : 295-300.

Chadha, K. L. 1992. The Indian orchid scenario. *J. Orchid Soc. India* 6 (1, 2) : 1-4.

Champagnat, M., G. Morel, and B. Mounetou. 1970. La multiplication vegitatire de *Cattleya* a partir de jeunes feuilleo cultivees aseptiquement *in vitro. Am. Sci. Nat. Bot.*, Ser. 12, 11 : 97-113.

Chaturvedi, H. C., and A. K. Sharma. 1986. Mericloning of orchids through culture of tips of leaves and roots. In : *Biology, conservation, and culture of orchids* (ed. S.P. Vij) pp. 469-72. Affiliated East-West Press, New Delhi.

Clement, E. 1924. The non-symbiotic germination of orchid seed. *Orchid Rev.* 32 (378) : 359-364.

Clement, E. 1925. A resume of the non-symbiotic germination of Orchid seed. *Orch. Rev.* 33 (385) : 199-200.

Cooper, A. 1983. *Paphiopedilum* mericlone project. Australian *Orchid Review*. pp. 20.

Corner, A. J., and M. D. Thomas. 1981. Re-establishing plantlets from tissue culture : a review, *Comb. Proc. Intern. Plant Prop Soc.* 31 : 342-357.

Curtis, J. T. 1943. Germination and seedling development in five species of *Cypripedium L. Amer. J. Bot.* 30 : 199-206.

Curtis, J. T., and Spoerl Edward. 1948. Studies on the nitrogen nutrition of orchid embryos. II. Comparative utilization of nitrate and ammonium nitrogen. *Am. Orchid Soc. Bull.* 17 : 111-114.

Dale, P. J. 1980. Embryoids from cultured immature embryos of *Lolium multiflorum. Z. Pflanzenphysiol.* 100 : 73-77.

Darwin, C. 1888. The various contrivances by which orchids are fertilized by insects. 2nd ed. John Murray, London, pp. 300.

Davies, W. E. 1970. White clover breedking - Review. In : *White clover Research Occasional Symp.,* No. 6 (J. Lowe, ed.) pp. 92-122. The British Grassland Soc.

de Bruijne, E., and P. Debergh. 1974. Response of *Cymbidium* protocorms to major element deficiency in a culture medium, Mededelingen Fakulteri Landbouw-wetenschappen. *Ghent* 39 : 210-15.

Downie, D. G. 1940. On the germination and growth of *Goodyera rapens. Trans. Bot. Soc. Edin.,* 33 : 36-51.

Downie, D.G. 1941. Notes on the germination of some British orchids. *Trans. and Proc. Bot. Soc. Edin.* 33 : 94-103.

Downie, D. G. 1943. Notes on the germination of *Corallorhiza innata. Trans. and Proc. Bot. Soc. Edin.,* 33 : 380-82.

Downie, D. G. 1949a. The germination of *Goodyera repens* (L.) R. Br. in fungal extract. *Trans. and Proc. Bot. Soc. Edin.* 35 : 120-125.

Downie, D. G. 1949 b. The germination of *Listera ovata* (L.) *Trans. and Proc. Bot. Soc.* Edinburgh, 35 : 126-130.

Duncan, R. E., and J. T. Curtis. 1942. Intermittent growth of fruits of *Cypripedium* and *Paphiopedilum :* A correlation of the growth of orchid fruits with their internal development. *Bull. Torrey Bot. Club,* 69 : 353-359.

Dunstan, D. I., and K. E. Turner. 1984. The acclimatization of micropropagated plants. In : Vasil I. K. (Ed.) *Cell Culture and Somatic Cell Genetics of Plants,* Vol. I. pp. 123-129. Academic Press, Orlando.

Dymes, T. A. 1923. Notes on the seeds of British Dactylorchids. *Bot. Soc. & Exch. Club,* 6 (3) : 462.

Ehrenberg, L. 1971. Higher plants. In : H. Hollaender (ed.) *Chemical Mutagens, principles and methods for their detection,* pp. 365-386, Plenum Press, New York.

Ernst, R. 1967. Effect of carbohydrate selection on the growth rate of freshly germinated *Phalaenopsis* and *Dendrobium* seed. *Amer. Orchid Soc. Bull.,* 36 : 1068-73.

Ernst, R. 1974. The use of activated charcoal in asymbiotic seedling culture of *Paphiopedilum. Amer. Orchid Soc. Bull.* 43 : 35-38.

Ernst, R. 1982. Seed germination and seedling culture—A Manual. In : *Orchid Biology : Reviews and Perspectives II* (Ed. J. Arditti) pp. 350-353. Cornell Univ. Press, Ithaca and London.

Ernst, R., and J. Arditti. 1990. Carbohydrate physiology of orchid seedlings. III. Hydrolysis of malto-oligosaccharides by *Phalaenopsis* (Orchidaceae) seedlings. *Amer. J. Bot.,* 77 (2) : 188-195.

Ernst, R., J. Arditti, and P. L. Healey. 1970. The nutrition of orchid seedlings. *Amer. Orchid Soc. Bull.,* 39 : 599-605.

Ernst, R., J. Arditti, and P. L. Healey. 1971. Carbohydrate physiology of orchid seedlings. Pt. 2. Hydrolysis and effect of oligosaccharides. *Amer. J. Bot.,* 58 : 827-35.

Fagundes, A. B. 1938. Germinacão asymbiotica das Sementes de Orchideas. *Orchidea* (Rio de Janeiro) 1 : 85-90.

Farrar, M. D. 1963. Growing orchid seedlings. *Amer. Orch. Soc. Bull.*, 32 (11) : 888-891.

Fast, G. 1974. Uber eine methode der kombinierte generativenvegetativen vermehrung von *Cypripedium calceolus* L. *Die orchidee*, 25 : 125-129.

Fast, G. 1976. Moglichkeiten zur Massenvermehrung von *Cypripedium calceolus* und anderen europaischen wild-orchideen, pp. 359-363. In : *Proc. 8th World Orchid Conf.*, Frankfurt.

Flamee, M. 1978. Influence of selected media and supplements on the germination and growth of *Paphiopedilum* seedlings. *Amer. Orchid Soc. Bull.*, 47 : 419-429.

Fonnesbech, M. 1972. Growth hormones and propagation of *Cymbidium in vitro. Physiol. Plant.*, 27 : 310-316.

Galzy, R. 1969. Recherches sur la croissance de *Vitis rupestris* scheele sain et court noue cultive *in vitro* a differentes-temperatures. *Ann. Phytopathol.* 1 : 149-166.

Gaul, H. 1964. Mutation in plant breeding. *Rad. Bot.*, 4 : 152-232.

Gautheret, R. J. 1985. History of plant tissue and cell culture : A personal account. In I. K. Vasil (Ed.), *Cell culture and somatic Cell Genetics of plants*, Vol. 2. Academic Press, New York, pp. 1-59.

Gavinlertvatana, P., and P. Prutpongse. 1991. Commercial micropropagation in Asia. In : *Micropropagation-Technology and Application* (eds. P. C. Debergh and R. N. Zimmerman) pp. 181-89. Kluwer Academic Press, Dordrecht.

Goh, C. J. 1970. Tissue culture of *Vanda* Miss Joaquim. *J. Singapore Nat. Acad. Sci.*, 2 : 31-33.

Goh, C. J. 1971. Some effects of auxin on orchid seed germination and seedling growth. pp. 65-66. In : International Symp. on Morphogenesis in Plant Cell, Tissue and Organ culture VIII. 4.

Goh, C. J. 1973. Meristem culture of *Aranda. Deborah. Malayan Orchid Rev.*, 12 : 10-13.

Goh, C. J. 1981. Development of Orchid Tissue Culture in South-East Asian countries. In : *Proc. Costed Symp. on Tissue culture of Economically Important plants.* (Ed. A. N. Rao), Singapore, pp. 205-209.

Goh, C. J., and Loh, C. S. 1975. Further observations on meristem culture of *Aranda. Deborah. Malayan Orchid* Rev., 12 : 10-13.

Goh, C. J., C. S. Loh, and A. N. Rao. 1975. Clonal propagation of *Aranda*, hybrids through shoot meristem culture. *Proc. of Seminar "Biology in Society"*, 26-35. Singapore Natl. Acad. Science and Singapore institute of Biology, Singapore.

Griffis, J. L., G. Hennen, R. P. Oglisby. 1983. Establishing tissue-cultured plant in soil. *Comb Proc. Intern Plant Prop. Soc.*, 33 : 618-622.

Guignard, L. 1886. Sur la pollinisation et see effets chez les Orchidées. *Ann. Sci. Nat. Bot.*, VII. 4 : 202-240.

Gunckel, J. E., and A. H. Sparrow. 1961. Ionizing radiation; Biochemical, Physiological and Morphological aspects of their effects on plants. In : W. Ruchland (ed.) XVI : *Encycl. Plant Physiol*, pp. 55-611, Springer-Verlag, Berlin.

Haber, A. H. 1968. Ionizing radiations as research tools. *A Rev. Physiol.*, 19 : 463-489.

Haberlandt, G. 1902. Kulturversuche mit isolierten Pflanznzellen. Sitzungser. Math-Naturwiss. *Kl. Kaiser Akad. Wiss. Wien* III : 69-92.

Hadley, G. 1970. The interaction of kinetin, auxin and other factors in the development of north temperate orchids. *New Phytol.*, 69 : 549-555.

Hadley, G., and G. Harvais, 1968. The effect of certain growth substances on asymbiotic germination and development of *Orchis purpurella. New Phytol.*, 67 : 441-45.

Hager, H. 1954. Growing Cattleyas from seed to flower in $2^1/_2$ years. *Amer. Orch. Soc.* Bull., 23 (2) : 78-81.

Handa, S. S. 1986. Orchids for Drugs and Chemicals. In : *Biology, Conservation and Culture of Orchids* (ed. S. P. Vij) Affiliated East West Press, New Delhi.

Harbeck, M. 1963. Einige Beobachtungen bei der Aussaat verschiedener europaischer Erdorchideen auf sterilem Nahrboden. Die *Orchidee* (Hamburg), 14 : 58-65.

Harley, J. L. 1951. Recent Progress in the study of endotrophic mycorrhiza (Abstract). *Amer. Orch. Soc. Bull.*, 20 (5).

Harn, C. 1968. Propagation of orchids and mutation breeding. *Amer. Orchid Soc. Bull.*, 37 : 1066-68.

Harvais, G. 1972. The development and growth requirements of *Dactylorhiza purpurella* in asymbiotic cultures. *Can. J. Bot.*, 50 : 1223-1229.

Harvais, G. 1973. Growth requirements and development of *Cypripedium reginae* in axenic culture. *Can J. Bot.*, 51 : 327-332.

Harvais, G. 1982. An improved culture medium for growing the orchid, *Cypripedium reginae* axenically. *Can J. Bot.*, 60 : 2547-55.

Hayes, A. B. 1969. Observations on orchid seed mycorrhizae. *Mycopathol. et Mycol. Appl.*, 38 : 139-144.

Hegarty, C. P. 1955. Observations on the germination of orchid seed. *Amer. Orchid Soc. Bull.*, 24 : 457-464.

Hew, C. S., and T. C. Mah. 1989. Sugar uptake and invertase activity in *Dendrobium. New Phytol.*, 111 : 167-171.

Hew, C. S., T. F. Chai, Y. K. Lee, and C. S. Loh. 1987. The need for a flow orchid tissue culture system. *Malayan Orchid Rev.*, 21 : 30-34.

Hirsh, D. H. 1959. Gibberellates : Stimulants for *Cattleya* seedlings in flasks. *Amer. Orch. Soc. Bull.*, 28 (5) : 342-344.

Hoene, F. C. 1940. Flora Brasilica. *Orchidaceas*. Fasc. 1, Vol. 12, pt. 1 : 1-12.

Hoene, F. C. 1949. Iconographia de Orchidaceas do Brazil. *Sicretaria de Agricultura,* Sao Paulo, 301 pp + 300 plates.

Holden, J. H. W., and J. T. Williams (eds.). 1984. Crop genetic resources, Conservation and Evaluation. Allen and Unwin, London.

Homés, J. 1973. Modification ultrastructuralis des chloroplastes de protocormes d'Orchidees cultives *in vitro* en presence de saccharose. *J. Microscopie,* 17 : 66a.

Homés, J., and N. Vansevéren-Van Espen. 1973. Effect du de saccharose et de la lumiere sur le diveloppement et la morphologie de protocormes d'orchidees cultives *in vitro. Bull. Soc. Roy. Bot. Belg.,* 106 : 89-106.

Homés, J. R. Freson, M. Vermylen, and M. Michel. 1971 a. Relation entre les conditions de culture et la morphogenese chez les protocromes d'orchidees cultives *in vitro. 96th Cong., Nat. Soc. Savantes,* Toulouse 4 : 93-105.

Homés, J., R. Freson., M. Vermylen, and M. Michel. 1971 b. Relation entre les conditions de culture et la morphogenese chez les protocromes d'Orchidees. p. 86. In : *96th Cong. Nat. Soc. Savantes,* Toulouse 4.

Huang, L. C. 1988. A procedure for asexual multiplication of Paphiopedilums *in vitro. Amer. Orchid Soc. Bull.,* 57 (3) : 274-278.

Hughes, K. W. 1981. Ornamental species. In : *Cloning Agricultural plants via in vitro Techniques* (B. V. Conger, ed.), pp. 5-50. CRC Press, Boca Raton, Florida.

Hyatt, D. W. 1965. Investigations in the Orchid family. *Amer. Orch. Soc. Bull.,* 34 (9) : 787-790.

Ilsley, P. 1966. Great Orchids. The Ilsley Orchids, Los Angeles, pp. 56.

Intuwong, O., and Y. Sagawa. 1973. Clonal propagation of Sarcanthine orchids by aseptic culture of inflorescences. *Amer. Orchid Soc. Bull.,* 42 : 209-215.

Intuwong, O., and Y. Sagawa. 1974. Clonal propagation of *Phalaenopsis* by shoot tip culture. *Amer. Orchid Soc. Bull.,* 43 : 893-895.

Intuwong, O., Kunisaki, T., and Y. Sagawa, 1972. Vegetative propagation of *Phalaenopsis* by flower stalk cuttings. N. A. Okika O Hawaii, 1 : 13-18.

Irawati, Harjadi, S. S., Suseno, H., and S. Idris. 1977. Tissue culture of *Aranthera* James Storie. *Orchid Rev.,* 85 : 138-142.

Israel, H. W. 1963. Production of *Dendrobium* seedlings by aseptic culture of excised ovularies. *Amer. Orch. Soc. Bull.,* 32 (6) : 441-443.

Ito, Itsuhiko, 1955. Germinations of seeds from immature pod and subsequent growth of seedlings in *Dendrobium nobile* Lindl. *Sci. Rpts. Saikyo Univ. of Ag.,* 7 : 35-42.

Jacques, H., Francoise, D., and B. Jean-louis. 1982. Cold storage of plant tissue cultures. *Proc. 5th Intt. Plant Tissue and Cell culture.* pp. 801-02.

Jain, S. K. 1986. Orchid wealth (?) of India. In : *Biology, Conservation and Culture of Orchids* (ed. S. P. Vij), Aff. East-West Press, N. Delhi. pp. 319-22.

Jain, S. K., and A. Mehrotra. 1984. A preliminary inventory of Orchidaceae in India. BSI, Howrah.

Jeyanayaghy, S., and A. N. Rao, 1966. Flower and seed development in *Bromheadia finlaysoniana*. *Bull. Torrey Botan. Club*, 93 : 97-103.

Kano, K. 1965. Studies on the media for orchid seed germination. *Mem. Fac. Agr. Kagawa Univ.*, 20 : 74.

Kartha, K. K. 1981. Meristem culture and Cryopreservation methods and applications. In : T. A. Thorpe (ed.), *Plant Tissue Culture.* Academic Press, N. York, pp. 181-211.

Karthikeyan, S., S. K. Jain, M. P. Nair, and Sanjappa. 1989. Florae Indicae Enumeratio : Monocotyledonae, BSI, Howrah, Ser. 4.

Katiyar, R. S., G. D. Sharma, and R. R. Mishra, 1986. Studies on mycorrhizal associations in terrestrial orchids. In : *Biology, Conservation and Culture of orchids* (ed. S.P. Vij). Aff. East-West Press, N. Delhi. pp. 63-70.

Kim, K.K., J.T. Kunisaki, and Y. Sagawa. 1970. Shoot tip culture of *Dendrobium. Amer. Orchid Soc. Bull.*, 39: 1077-1080.

Knudson, L. 1921. La germinacion no simbiotica de las semillas de orquideas. *Bol. Real. Soc. Española Hist. Nat.*, 21: 250-260.

Knudson, L. 1922. Non-symbiotic germination of orchid seeds. *Bot. Gaz.*, 73: 1-25.

Knudson, L. 1924. Further observations on non-symbiotic germination of orchid seeds. *Bot. Gaz.*, 77: 212-19.

Knudson, L. 1925. Physiological study of the symbiotic germination of orchid seeds. *Bot. Gaz.*, 79 (4): 345-379.

Knudson, L. 1929. Physiological investigations on orchid seed germination. *Proc. Int. Cong. Plant Sci.*, 2: 1183-1189.

Knudson, L. 1930. Flower production by orchids growth non symbiotically. Bot. Gaz., 89: 192-199.

Knudson, L. 1946. A new nutrient solution for germination of orchid seed. *Amer. Orchid Soc. Bull.*, 15: 214-217.

Knudson, L. 1950. Germination of seeds of Vanilla. Amer. J. Bot., 37: 241-247.

Knudson, L. 1951. Nutrient solutions for orchids. Bot. Gaz., 112 (4): 528-532.

Kofranek, A. 1957. Some notes on orchids. Unpublished mimeograph, Dept. of Floriculture, Univ. Calif. (Los Angeles).

Kotomori, S., and T. Murashige. 1965. Some aspects of aseptic propagation of orchids. *Amer. Orchid Soc. Bull.*, 34: 484-489.

Kriechbaum. 1953. Nahrboden für Orchideeanzucht aus Samen. *Garten Zeitschrift Illustrierte Flora*, 76(2): 17-18.

Krikorian, A.D., and D.L. Berquam. 1969. Plant cell and tissue cultures: The role of Haberlandt. Botan. Rev., 35: 59-67.

Kukulcznaka, K., Kromer, K., B. Klonowska, and B. Czestka. 1990. Effect of temperature on long term storage of Droseraceae plant cultures in *in vitro* conditions.

Kumar, S.P.K. 1992. Potting media and post-transplantation growth of *Dendrobium* hybrid seedlings. *J. Orchid Soc. India,* 6(1,2): 131-133.

Kunisaki, J.T., K.K. Kim, and Y. Sagawa. 1972. Shoot tip culture of *Vanda*. *Amer. Orchid Soc. Bull.,* 41: 435-439.

Kupper, W., and W. Linsenmaier. 1961. Orchids. Thomas Nelson and Sons Ltd., London, pp. 128.

La Garde, R.V. 1929. Non-symbiotic germination of orchids. *Ann. Missouri Bot. Gard.,* 16(4) : 499-514.

Lawrence, D., and J. Arditti, 1964. A new medium for the germination of orchid seeds. *Orch. Soc. Southern California Rev.,* 6(8): 4-5; also *Amer. Orch. Soc. Bull.,* 33(9) : 766-68.

Leach, G.N. 1979. Growth in soil of plantlets produced by tissue culture, loblolly pine. *Tappi* 62: 59-61.

Letham, D.S. 1974. Regulators of cell division in plant tissues XX. The cytokinins of coconut milk. *Physiol. Plant.,* 32: 66-70.

Lindemann, E.G.P. 1967. Growth requirements for meristem culture of *Cattleya. Diss. Abstr. Sut.* B. 28: 2284-2285.

Linden, B. 1980. Aseptic germination of seeds of northern terrestrial orchids. *Ann. Bot. Gennici,* 17: 174-82.

Loh, C.S., A.N. Rao, and Goh, C.J. 1975. Clonal propagation from leaves in the orchid *Aranda*. *Jour. Singapore Nat. Acad. Sci.,* 4: 97-99.

Lugo, H.L. 1955. The effect of nitrogen on the germination of *Vanilla planiflora. Amer. J. Bot.,* 42: 679-84.

Lundergan, C., and J. Janick. 1979. Low temperature storage of *in vitro* apple shoots. *Hortscience,* 14: 514.

Magrou, J. 1944. Sur le rôle des mycorrhizes endotrophes. *Compt. Rend. Acad. Sci.,* Paris 219: 519-521.

Maheshwari, P., and S. Narayanswami. 1952. Embryological studies on *Spiranthes australis* Lindl. *Jour. Linn. Soc.* London (Bot.), 53: 474-486.

Marston, M.E. 1966. Vegetative propagation of plants using tissue culture techniques. Nottingham Univ. *School of Agri. Report* 77-80.

Marston, M.E. 1969. Vegetative propagation *in vitro*, pp. 339-394. In: W.J. Whittington (ed.), Root growth, Butterworth, London.

Marston, M.E., and P. Voraurai. 1967. Multiplication of orchid clones by shoot meristem culture. A review of the literature. Univ. Nottingham. Dept. Hort., *Misc. Pub.,* 17: 1-18.

Mathews, V.H., and P.S. Rao. 1985. *In vitro* culture of *Vanda* Hybrid (Vanda TMA x Vanda Miss Joaquim) I. Studies on protocorm explants. In: *Proc. Indian Nat. Sci. Acad.* B 51 N.1, pp. 96-103.

Mc. Ewan, B. 1961. Stock solution medium for flasking. *Orch. Soc. Southern California Rev.,* 3(5) : 7-9.

Mead, J.W., and C. Bullard. 1975. Effects of vitamins and nitrogen sources on symbiotic germination and development *Orchis laxiflora* and *Ophrys sphegodes*. *New Phytol.*, 74: 33-40.

Mead, J.W., and C. Bullard. 1979. Vitamins and nitrogen requirements of *Orchis laxiflora* Lamk. *New Phytol.*, 83: 129-136.

Meyer, J.R. 1944. Meio assimbiotieo simplificado para culturas de sementes de orquideas. *O'Biologico*, 10(3): 63-65.

Mitra, G.C. 1986. *In vitro* culture of orchid seeds for obtaining seedlings. In: *Biology, Conservation, and Orchid Culture* (ed. S.P. Vij), pp. 401-412.

Miyazaki, S. 1968. On the germination and growth of the Orchid seeds irradiated by gamma rays. *Agric. Bull. Saga Univ.*, 43: 101-116.

Miyazaki, S., and T. Nagamatsu. 1965. Studies on the promotion of the early growth *in vitro* of orchid, Pt. I. *Agric. Bull. Saga Univ.*, 21: 131-149.

Mohammed, H.G., and W.E. Vidaver. 1988. Root Production and Plantlet Development in tissue-cultured conifers. *Plant Cell, Tissue and Organ Culture*, 14: 107-160.

Morel, G. 1960. Producing virus-free *Cymbidiums*. *Amer. Orchid Soc. Bull.*, 29: 495-497.

Morel, G. 1964. Tissue culture—A new means of clonal propagation of orchids. *Amer. Orch. Soc. Bull.*, 33: 473-478.

Morel, G. 1965 a. Clonal propagation of orchids by meristem culture. *Cymbidium Soc. News*, 20: 3-11.

Morel, G. 1965 b. Eine neue methode erbgleicher vermehrung: die Kulturron Triebspitzen-Meristememn. *Die Orchidee*, 16: 165-176.

Morel, G. 1974. Clonal multiplication of orchids, pp. 169-222. In: C.L. Withner (ed.) *The Orchids: Scientfic Studies*, Wiley-Interscience, New York.

Mullin, R.H., and D.E. Schlegel. 1976. Cold storage maintenance of strawberry meristem plantlets. *Hortscience*, 11: 100-101.

Murashige, T. 1974. Plant propagation through tissue culture. *Ann. Rev. Plant Physiol.*, 25: 135-66.

Murashige, T. 1977. Manipulation of organ initiation in plant tissue cultures. *Bot. Bull. Acad. Sinica.*, 18: 1-24.

Murashige, T., and F. Skoog. 1962. A revised medium for rapid growth and bioassays with tobacco tissue cultures. *Physiol. Plant.*, 473-97.

Nagashima, T. 1982. Studies on the seed germination and embryogenesis in the *Cymbidium goeringii* Rchb. f. and *Paphiopedilum insigne* var. *sanderae* Rchb. f. *J. Japan Soc. Hort. Sci.*, 51(1): 94-105.

Nakamura, S.J. 1982. Nutritional conditions required for the non-symbiotic culture of an achlorophyllous orchid *Galeola septentrionalis*. *New Phytol.*, 90: 701-15.

Ng, S.Y.C., and N.Q. Ng. 1991. Reduced growth storage of germplasm. In : J.H. Dodds (ed.), In: *in vitro methods for conservation of plant genetic resources*. Chapman and Hall, London, pp. 11-40.

Nimoto, D.H., and Y. Sagawa. 1961. Ovule development in *Dendrobium*. *Amer. Orchid Soc. Bull.*, 30: 813-19.

Oliva, A.P., and J. Arditti. 1984. Seed germination of North American orchids. II. Native California and related species of *Aplectrum, Cypripedium and Spiranthes*. *Bot. Gaz.*, 145: 1495-501.

Ong, P.Y. 1969. Physiological studies on pod development and seed germination of *Vanda* Miss Joaquim. *Honors Res. Enercise*, Univ. Singapore.

Ordin, L., and J. Bonner. 1957. Effect of galactose on growth and metabolism of *Avena* coleoptile sections, *Plant Physiol.*, 32: 212-215.

Pastrana, M.D., and J.K. Santos. 1931. A contribution on the life history of *Dendrobium anosmum* Lindley. *Nat. and App. Sci. Bull.* Philippines Univ. 1: 133-44.

Pathak, P., S.P. Vij, and K.C. Mahant. 1992. Ovule Culture in *Goodyera biflora* (Lindl.) Hk. F. : A study *in vitro*. *J. Orchid Soc. India*, 6(1, 2) : 49-53.

Perik, R.L.M., and H.H.M. Steegmans. 1972. The effect of 6-bezylamino purine on growth and development of *Cattleya* seedlings grown unripe seeds. *Z. Pflanzenphysiol.*, 68: 228-234.

Poddubnaya-Arnoldi, V.A. 1960. Study of fertilization in the living material of some angiosperms. *Phytomorph.*, 10: 185-198.

Poddubnaya-Arnoldi, V.A., and A. Selezneva. 1957. Orchidei in koultoura (Orchids and their culture). *Acad. Sci.*, U.S.S.R. Moscow. 256 pp.

Quednow, K.G. 1930. Beiträge zur Frage der Aufnahme geloster Kohleenstoffverbindungen bei Orchideen und andere Pflanzen. *Bot. Archiv.*, 30: 51-108.

Raghavan, V. 1964. Effects of certain organic nitrogen compounds on growing *in vitro* of seedlings of *Cattleya*. *Bot. Gaz.*, 125: 260-67.

Raghavan, V. 1976 . Experimental embryogenesis in vacular plants. Academic Press, New York.

Raghavan, V., and J.G. Torrey. 1964. Inorganic nitrogen nutrition of the embryos of the seedlings of the orchid *Cattleya*. *Amer. J. Bot.*, 51: 264-74.

Raghuwanshi, A.N., R.R. Mishra, and G.D. Sharma. 1985. Effect of synthetic media on Asymbiotic seed Germination and Seedling Growth of *Dendrobium noblle* and *Sarcanthus pallidus*. *Proc. Indian Natn. Sci. Acad.*, B. 51, No. 3 pp. 360-363.

Raghuwanshi, A.N., R.R. Mishra, and G.D. Sharma. 1986. Effect of pH on Asymbiotic seed germination and seedling development of orchid. In: *Biology, Conservation and Culture of orchids* (ed. S.P. Vij). Affiliated East West Press, N. Delhi.

Rao, A.N. 1977. Tissue culture in the orchid industry. In: *Applied and fundamental aspects of plant cell, tissue and organ culture* (eds. J. Reinert and Y.P.S. Bajaj). Springer-Verlag, Berlin.

Rao, A.S. 1979. Orchids of India. National Book Trust, N. Delhi.

Rao, A.N., and P.N. Avadhani. 1964. Some aspects of *in vitro* culture of *Vanda* seeds. Proc. *4th World Orchid Conf.*, Singapore. pp. 194-202.

Reilly, K. Washer, J. 1977. Vegetative propagation of radiata pine by tissue culture: plantlet formation from Embryonic tissue. *NZJ. For. Sci.*, 7(2): 199-206.

Robinson, J. 1941. Nutritional studies with orchids. *Amer. Orch. Soc. Bull.*, 10(7): 185-88.

Rolfe, R.A. 1962. Seeds and Pollination. *Orch. Rev.*, 70 (831): 282-83.

Rotor, G., Jr. 1949. A method for vegetative propagation of *Phalaenopsis* species and hybrids. *Amer. Orchid Soc. Bull.*, 18: 738-39.

Sagawa, Y., and J.T. Kunisaki. 1984. Clonal propagation: orchids. In: Vasil, I.K. (ed.), *Cell culture and somatic cell genetics of plants,* Vol, 1, pp. 61-67. Academic Press., New York.

Salisbury, R.A. 1804. On the germination of the seeds of orchidaceae, *Trans. Linn. Soc.* London. 7: 29-32.

Salisbury, E.J. 1942. The reproductive capacity of plants. G. Bell and Sons Ltd., London.

Sampolinski, J. 1965. Adventures in orchid growing. *Amer. Orch. Soc. Bull.*, 34(7) : 602-604.

Schaffstein, G. 1938. Untersuchungen uber die Avitaminose der Orchideen Keimlinge. *Jahrb. Wiss. Bot.*, 86: 720-752.

Scott, R.J., and J. Arditti. 1959. *Cymbidium* from pod to pot. *Amer. Orch. Soc. Bull.*, 28(11): 823-29.

Scowcroft, W.R., S.A. Ryan, R.I.S. Brettell, and P.J. Larkin. 1985. Biotechnology in international agricultural research. IRRI Press, Manila, pp. 99.

Scully, R.M., Jr. 1967. Aspects of meristem culture in the *Cattleya* alliance. *Amer. Orchid Soc. Bull.*, 36: 103-108.

Seeman, G. 1953 a. Orchideenanzucht aus Samen Gart. *Zeitschr. Illus. Flora* 76(2): 15-17.

Seeman, G. 1953 b. Uber eine neue Wasserungs methode fur schwerkeimende Orchideensamen. *Die Orchidee,* 4(2).

Seeni, S. 1988. Micropropagation of blue *Vanda* using leaf bases. In: *Proc. Natl. Seminar on Current Research Trends in Indian Orchids with a special reference to tissue culture technology.* pp. 22. NEHU, Shillong.

Seeni, S., and P.G. Latha. 1990. Post-transplantation growth of *Phalaenopsis* hybrid seedlings in community pots. *J. Orchid Soc. India,* 4 (1,2): 127-33.

Shantz, E.M., and F.C. Steward. 1964. Growth promoting substances from the environment of the embryo II. The growth stimulating complexes of coconut milk, corn and Aessulus. In: *Regulateurs naturels de la croissance* vegetable collog Intern Centre Nat. Rech. France: Gif sur-yvette, pp. 59-75.

Sharma, S.K., and P. Tandon. 1990. Asymbiotic germination and seedling growth of *Cymbidium elegans* Lindl. and *Coelogyne punctulata* Lindl. as influenced by different carbon sources. *J. Orchid Soc. India*, 4(1,2): 149-160.

Sharon, M., and G. Vasundhara. 1990. Micropropagation of *Dendrobium* Joannie Ostenhault. *J. Orchid Soc. India*, 4 (1, 2) : 145-148.

Shushan, S. 1959. Developmental anatomy of an orchid *Cattleya* x *Trimos*. In: *The orchids: A scientific survey* (ed. C.L. Withner) Ronald Press, N. York.

Sideris, C.P. 1950. A nutrient solution for germination of orchid seeds. *Bull. Pac. Orchid Soc.*, Hawaii, 8: 337-39.

Singh, H. 1976. Meristem tissue culture of *Dendrobium*. Ng Eng Cheow. *Gardners' Bull.*, Singapore, 28: 259-67.

Singh, F. 1992. *In vitro* propagation of orchids 'state of the art'. *J. Orchid Soc. India*, 6(1, 2): 11-14.

Skoog, F., and C.O. Miller. 1957. Chemical regulation of growth and organ formation in plant tissues cultivated *in vitro*. In: *Biological Action of Growth Substances. 11th Symp. Soc. Exp. Biol.*, 11: 118-31.

Smith, F. 1932. Raising orchid seedlings asymbiotically. *Gard. Chron.*, 91 (2349): 9-11.

Sood, A., and S.P. Vij. 1986. *In vitro* root segment culture of *Rhynchostylis retusa* Bl. In: *Biology, Conservation and Culture of orchids* (ed. S.P. Vij), pp. 463-68. Aff. East-West Press, N. Delhi.

Steward, F.C. 1961. Growth induction in explant cells and tissues metabolic and morphogenetic manifestation. In: Rudnik D. (ed.), *Synthesis of moledular and cellular structure. 19th Symp. Soc. Study development and growth*, pp. 193-246.

Sparrow, A.H. 1960. Uses of large sources of ionizing radiations in botanical research and some possible practical applications. In: *Large Radiation sources in Industry*, 2: 195-219. I.A.E.A. Vienna.

Spoerl, E. 1948. Amino acids as sources of nitrogen for orchid embryos. *Amer. J. Bot.*, 35: 88-95.

Steward, F.C., and M.O. Mapes. 1971. Morphogenesis in aseptic cell cultures of *Cymbidium*. *Bot Gaz.*. 132: 65-70.

Stoutamire, W. 1974. Terrestrial orchid seedling. pp. 101-128. In: C.L. Withner (ed.), *The orchids: Scientific studies*. Wiley-Interscience, N. York.

Stoutamire, W. 1981. Early growth in North American terrestrial orchid seedlings. In: *Proc. Symp. II and Lectures. North American Terrestrial orchids*, pp. 14-24 (ed. E.H. Planton).

Strauss, M.S., and D.M. Reisinger. 1976. Effects of napthaleneacetic acid on seed germination. *Amer. Orchid Soc. Bull.*, 45: 722-723.

Sun, C.N. 1955. Anomalous Structure in the hypocotyl of Soybean roots by 2,4-dichlorophenoxyacetic acid. *Science*, 121: 641.

Swamy, B.G.L. 1949. Embryological studies in the Orchidaceae. II. Embryogeny. *Am. Mide. Nat.*, 41: 202-32.

Tamanaha, L.R., C.G. Shimizu and J. Arditti. 1979. The effects of ethephon on *Cattleya auramtiaca* (Orchidaceae) seedlings. *Bot Gaz.*, 40: 25-28.

Tanaka, M., K. Kumura, and M. Goi, 1988. Optimal conditions for shoot production from *Phalaenopsis* flower stalk cuttings cultured *in vitro. Scientia Horticulturae,* 35: 117-26.

Teo, Ç.K.H., J.T. Kunisaki, and Y. Sagawa. 1973. Clonal propagation of strap-leafed *Vanda* by shoot-tip culture. *Amer. Orch. Soc. Bull.*, 42: 402-405.

Thimann, K.V. 1956. Studies on the growth and inhibition of isolated plant parts. V. The effects of Cobalt and other metals. *Amer. J. Bot.*, 43: 241-50.

Thomale, H. 1954. Die Orchideen. Eugen Ulmer, Stuttgart., p. 189.

Thorpe, T.A. 1990. The current status of plant tissue culture. In: *Plant Tissue Culture: Applications and Limitations* (ed. S.S. Bhojwani), Elsevier, Netherland.

Tomes, D.I. 1979. A tissue culture procedure for propagation and maintenance of *Lotus corniculatus* genotypeds. *Can. J. Bot.*, 57: pp. 137-40.

Tonnier, J. P. 1954. Mise au point d'un milieu de semis pour les graines d'une espéce de Vanillier dont on nóbtient pas la germination sur les milieux de culture classiques. *VIIIe Congr. Int. Bot.*, sec., 11-12, pp. 411-412.

Tournay, R. 1960. Orchidaceae Africanae. III. Le nombre de graines d'une capsule *d'Eulophia horsfallii* (Batem) *Summ. Bull. Jard. Bot. Etat* (Belgium) 30(4): 407-10.

Treub, M. 1879. Notes sur l'embryogénie des quelques orchidees. Verh. Akad. Amsterdam, Natuurk.

Tse, A.T.-Y., R.J. Smith, and W.P. Hackett. 1971. Adventitious shoot formation on *Phalaenopsis* nodes. *Amer. Orchid Soc. Bull.*, 40: 807-10.

Vacin, E.F. 1950. Some problems of germinating of *Cymbidium* seeds and growing seedlings. *Cymbidium Soc. News.* 5(2): 8.

Vacin, E., and F.W. Went. 1949. Some pH changes in nutrient solutions. *Bot. Gaz.*, 110: 605-613.

Vajrabhaya, T. 1977. Variation in clonal propagation 178-209. In: Arditti, J. (ed.) *Orchid Biology: Reviews and Perspectives*, 1. Cornell Univ. Press, Ithaca.

Vajrabhaya, M., and T. Vajrabhaya. 1970. Tissue culture of *Rhynchostylis gigantea*, a monopodial orchid. *Amer. Orchid Soc. Bull.*, 39: 907-10.

Van Staden, J., and Drewes, S.E. 1975. Identification of Zeation and Zeation—riboside in coconut milk. *Physiol. Plant.*, 34: 106-109.

Van Waes, J.M., and Debergh, P.C. 1986. *In vitro* germination of some western European orchids. *Physiol. Plant.*, 67: 253-61.

Vermeulen, P. 1947. Studies on *Dactylorchis.* Drukkhrij Fa Schotamus and Jens, Utrecht.

Vij, S.P., and P. Pathak. 1988. Asymbiotic germination of the saprophytic orchid, *Cymbidium macrorhizon:* a study *in vitro. J. Orchid Soc. India,* 2 (1, 2): 25-32.

Vij, S.P., and P. Pathak. 1990. Micropropagation of orchids through leaf segments. *J. Orchid Soc. India,* 4: 69-88.

Vij, S.P., A. Sood and K.K. Plaha. 1981. *In vitro* seed germination of some epiphytic orchids. In: *Contemporary trends in Plant Sciences* (ed. S.C. Verma) Kalyani Publishers, N. Delhi.

Vij, S.P., A. Sood, M. Sharma, and K.K. Plaha. 1984. Propagation of *Rhynchostylis retusa* Bl. (Orchidaceae) by direct organogenesis from leaf segments. *Bot. Gaz.,* 145(2): 210-14.

Voth, W. 1976. Aussaat und Kultur von *Serapias parviflora* und *S. orientalis,* pp. 351-58. In: *Proc. 8th World Orchid Conf.,* Frankfurt, (1975).

Werblin, T.P. 1963. Advanced microscopy and the orchid. *Orch. Dig.,* 27(8) : 340-42.

Werckmeister, P. 1970. Uber die Lichtinduktion der geotropen Orientierung von Luft-und Bodenwurzeln in Gewebekulturen von *Cymbidium. Ber. Dtsch. Bot. Ges.,* 83: 19-26.

Werckmeister, P. 1971. Light induction of geotropism and the contorl of proliferation and growth of *Cymbidium* in tissue culture. *Bot. Gaz.,* 132: 346-50.

White, P.R. 1934. Potentially unlimited growth of excised tomato root tips in a liquid medium. *Plant Physiol.* 9: 585-600.

Wilfret, G.J. 1966. Formation of protocorm-like bodies on excised *Cymbidium* shoot tips. *Amer. Orchid Soc., Bull,,* 35: 823-27.

Wilkes, G. 1983. CRC *Crit. Rev.,* 1: 131.

Wimber, D.E. 1963. Clonal multiplication of Cymbidiums through tissue culture of the shoot meristem. *Amer. Orchid Soc., Bull.,* 32: 105-107.

Wirth, M., and C.L. Withner. 1959. Embryology and development in the Orchidaceae. In: *The orchids: A scientific survey* (ed. C.L. Withner). Ronald Press, N. York.

Withers, L.A. 1985, Cryopreservation and storage of germplasm. *Plant cell culture-a practical approach* (ed. R.A. Dixon), pp. 169-191.

Withner, C.L. 1953. Germination of "Cyps". *Orchid Jour.,* 2: 473-77.

Withner, C.L. 1955. Ovule culture and growth of *Vanilla* seedlings. *Amer. Orchid Soc. Bull.,* 24: 380-92.

Withner, C.L. 1959. Orchid physiology, pp. 315-60. In: C.L. Withner (ed.) *The Orchids,* Ronald Press, N. York.

Withner, C.L. 1974. Development in orchid physiology. pp. 129-168. In: C.L. Withner (ed.), *The Orchids: Scientific Studies.* Wiley-Interscience, N. York.

Wolff, J. 1923. Contribution á la connassiance des phenoménes de symbiose chez les orchidees. *Compt. Rend. Acad. Sci.,* Paris, 177: 554-55.

Wolter, K.E. 1968. Root and shoot initiation in aspen callus culture. *Nature,* 219: 509-10.

Wolter, K.E., and F. Skoog. 1966. Nutritional requirements of *Fraxinus* callus culture. *Am. J. Bot.,* 53 (3): 262-69.

Wynd, F.L. 1933. Sources of carbohydrate for germination and growth of orchid seedlings. *Ann. Missouri Bot. Gard.,* 20: 569-81.

Yam, T.W., and M.A. Weatherhead. 1988. Germination and seedling development of some Hong-kong Orchids I. Lindleyana 3: 156-160.

Yamada, M. 1963 a. A simplified method of orchid seed germination. *Hawaiian orchid Soc. Inc.,* Honolulu, Mimeograph.

Yamada, M. 1963 b. Personal communication.

Zimmer, K., and W. Peipers. 1975.Weitre Untersuchungen zur kultur *vitro*. Acchmea gartenban wissenschaft, 40: 129-132.

Index